Nuclear Power and the Environment

ISSUES IN ENVIRONMENTAL SCIENCE AND TECHNOLOGY

TITLES IN THE SERIES:

How to obtain future titles on publication

A subscription is available for this series. This will bring delivery of each new volume immediately on publication and also provide you with online access to each title via the Internet. For further information visit http://www.rsc.org/issues or write to the address below.

For further information please contact:
Sales and Customer Care, Royal Society of Chemistry, Thomas Graham House, Science Park, Milton Road, Cambridge, CB4 0WF, UK
Telephone: +44 (0)1223 432360, Fax: +44 (0)1223 426017, Email: sales@rsc.org

ISSUES IN ENVIRONMENTAL SCIENCE AND TECHNOLOGY

EDITORS: R.E. HESTER AND R.M. HARRISON

32
Nuclear Power and the Environment

RSC Publishing

ISBN: 978-1-84973-194-2
ISSN: 1350-7583

A catalogue record for this book is available from the British Library

Published by The Royal Society of Chemistry,
Thomas Graham House, Science Park, Milton Road,
Cambridge CB4 0WF, UK

Registered Charity Number 207890

For further information see our web site at www.rsc.org

Preface

It is often forgotten that the human species has evolved in an environment surrounded by natural sources of radioactivity. These sources are as diverse as cosmic ray particles from space, potassium-40 in igneous rocks and the radioactive decay products of radon, a gas emanating from the land beneath our feet. Indeed, for many people, the so-called "radon daughters" pose the largest health risk incurred by breathing air indoors. However, in the public mind, artificial radioactivity is far more important, and since the cessation of atmospheric nuclear weapon testing this is primarily the radiation associated with the nuclear fuel cycle. This has already caused major pollution issues and continues to have the potential to do so, unless handled with great competence. The early years of nuclear power saw great enthusiasm for building nuclear power stations stimulated by lavish but unfulfilled promises of cheap electricity for all. This was followed by a period of disillusionment as the true costs of building nuclear power stations, generating the power and subsequent decommissioning became fully recognised, and the majority of investment went into fossil fuel sources of power. However, recent years have seen an increasing acceptance by politicians and the general public of the inevitability of damaging levels of climate change unless greenhouse gas emissions are curbed, and one of the few effective ways of doing so is through the adoption of nuclear power as a primary means of energy generation.

This volume is designed to provide an overview of some of the most important aspects of this field of science. In the first chapter, John Walls maps the technical and societal context in which nuclear power has existed since the first construction of experimental reactors. This serves to highlight many of the important issues which are taken up in later chapters, including issues such as the availability of uranium as a nuclear fuel, the consequences of fuel reprocessing, the economics of power generation and the costs of decommissioning. Other issues not explored elsewhere in depth within the volume, such as public attitudes to nuclear power, are also touched upon. The issue of nuclear fuel cycles and their by-products and consequences

Issues in Environmental Science and Technology, 32
Nuclear Power and the Environment
Edited by R.E. Hester and R.M. Harrison

Published by the Royal Society of Chemistry, www.rsc.org

for the environment are expanded upon in the second chapter by Francis Livens, Clint Sharrad and Laurence Harwood. In particular, this chapter highlights the limitations posed by the availability of uranium as a fuel, and the advantages and disadvantages of fuel reprocessing. The latter was developed originally largely to generate plutonium for military purposes but has gained a rather poor reputation because of discharges to the environment, and most countries now plan to store rather than reprocess spent fuel.

One of the major drivers of public opinion on nuclear power is the occurrence of nuclear accidents. Some, such as Windscale, Three Mile Island and Chernobyl, are well known to all but others occurring in the former Soviet Union were kept secret from the general public, yet have generated contamination which persists to this day. In the third chapter, Jim Smith describes the causes and implications of these accidents and puts the topic into context. Both major and minor accidents and planned releases of radioactive materials have led to land contamination and have generated low-level wastes which need to be stored safely. In the fourth chapter, Jon Lloyd, Francis Livens and Rick Kimber outline the issues raised by such contamination and describe some of the consequences and the available remediation techniques. Perhaps the greatest Achilles' heal of nuclear power generation is the fact that decommissioning of nuclear sites is required at the end of their active life, although interim "storage" may be used to allow cooling of the radioactivity by decay of the shorter-lived radionuclides. In Chapter 5, Anthony Banford and Richard Jarvis describe the legacy of contaminated nuclear sites and the approaches taken towards decontamination, and their positive and negative attributes.

The sixth chapter, by Katherine Morris, Gareth Law and Nick Bryan, deals with the geological disposal of higher activity wastes. This is currently a topical issue for many countries who have declared policies of constructing deep geological repositories for high and intermediate level waste with a view to safe storage on a timescale of at least a million years. The many considerations which go into the siting and design of such a repository are considered in this chapter. In the seventh chapter, by Joanna Renshaw, Stephanie Handley-Sidhu and Diana Brookshaw, the pathways of radioactive substances in the environment are described. This highlights how the chemistry of the actinides and fission products determines their behaviour in the environment which, in turn, influences their mobility and ultimate potential to cause exposure of humans and other biota. Chapter 8, by Brenda Howard and Nick Beresford, describes how radioactive substances translocate into biological organisms and the resultant dosimetry, and in Chapter 9, Richard (Jan) Pentreath describes the human consequences of exposure to environmental radioactivity. For many years, radiological protection was based upon the concept that measures adequate to protect human health would also be protective of the non-human biota. This paradigm has now shifted to one in which assessments are made of the dose to representative animals and plants and the likely consequences of those doses.

Overall, the volume provides a selective but broad overview of current issues in this long-standing but increasingly topical field, which we believe will be of immediate and lasting value, not only to practitioners in government, consultancy and industry but also to environmentalists, policy-makers and students taking courses in environmental science, engineering and management.

After the receipt of chapters from authors but before proof correction, the Japanese tsunami caused damage to the Fukushima nuclear plant which went into partial meltdown. At the time of production of this volume this situation was continuing with very little definitive information available. Where possible, authors have included this in their chapters but it is clear that a more complete view of the incident will only emerge well after the production of this book.

Ronald E. Hester
Roy M. Harrison

Contents

Issues in Environmental Science and Technology, 32
Nuclear Power and the Environment
Edited by R.E. Hester and R.M. Harrison

Published by the Royal Society of Chemistry, www.rsc.org

Editors

Ronald E. Hester, BSc, DSc (London), PhD (Cornell), FRSC, CChem

Ronald E. Hester is now Emeritus Professor of Chemistry in the University of York. He was for short periods a research fellow in Cambridge and an assistant professor at Cornell before being appointed to a lectureship in chemistry in York in 1965. He was a full professor in York from 1983 to 2001. His more than 300 publications are mainly in the area of vibrational spectroscopy, latterly focusing on time-resolved studies of photoreaction intermediates and on biomolecular systems in solution. He is active in environmental chemistry and is a founder member and former chairman of the Environment Group of the Royal Society of Chemistry and editor of 'Industry and the Environment in Perspective' (RSC, 1983) and 'Understanding Our Environment' (RSC, 1986). As a member of the Council of the UK Science and Engineering Research Council and several of its sub-committees, panels and boards, he has been heavily involved in national science policy and administration. He was, from 1991 to 1993, a member of the UK Department of the Environment Advisory Committee on Hazardous Substances and from 1995 to 2000 was a member of the Publications and Information Board of the Royal Society of Chemistry.

Roy M. Harrison, BSc, PhD, DSc (Birmingham), FRSC, CChem, FRMetS, Hon MFPH, Hon FFOM

Roy M. Harrison is Queen Elizabeth II Birmingham Centenary Professor of Environmental Health in the University of Birmingham. He was previously Lecturer in Environmental Sciences at the University of Lancaster and Reader and Director of the Institute of Aerosol Science at the University of Essex. His more than 350 publications are mainly in the field of environmental chemistry, although his current work includes studies of human health impacts of atmospheric pollutants as well as research into the chemistry of pollution phenomena. He is a past Chairman of the Environment Group of the Royal Society of Chemistry for whom he has edited 'Pollution: Causes, Effects and Control' (RSC, 1983; Fourth Edition, 2001) and 'Understanding our Environment: An Introduction to Environmental Chemistry and Pollution' (RSC, Third Edition, 1999). He has a close interest in scientific and policy aspects of air pollution, having been Chairman of the Department of Environment Quality of Urban Air Review Group and the DETR Atmospheric Particles Expert Group. He is currently a member of the DEFRA Air Quality Expert Group, the DEFRA Expert Panel on Air Quality Standards, and the Department of Health Committee on the Medical Effects of Air Pollutants.

List of Contributors

Anthony Banford, *National Nuclear Laboratory, Chadwick House, Risley, Warrington, Cheshire WA3 6AE, United Kingdom*

Nick Beresford, *Centre for Ecology & Hydrology, Lancaster Environment Centre, Library Avenue, Bailrigg, Lancaster LA1 4AP*

Diana R. Brookshaw, *School of Earth, Atmospheric and Environmental Sciences, University of Manchester, Manchester, M13 9PL, United Kingdom*

Nick Bryan, *Centre for Radiochemistry Research, School of Chemistry, The University of Manchester, Manchester, M13 9PL, United Kingdom*

Stephanie Handley-Sidhu, *School of Geography, Earth & Environmental Sciences, University of Birmingham, Edgbaston, Birmingham B15 2TT, United Kingdom*

Laurence Harwood, *Department of Chemistry, University of Reading, Whiteknights, Reading RG6 6AD, United Kingdom*

Brenda Howard MBE, *Centre for Ecology & Hydrology, Lancaster Environment Centre, Library Avenue, Bailrigg, Lancaster LA1 4AP*

Richard Jarvis, *National Nuclear Laboratory, Chadwick House, Risley Warrington WA3 6AE, United Kingdom*

Rick Kimber, *School of Earth, Atmospheric and Environmental Sciences, Williamson Building, The University of Manchester, Sackville Street, Manchester M13 9PL, United Kingdom*

Gareth Law, *School of Earth, Atmospheric and Environmental Sciences, The University of Manchester, Sackville Street, Manchester M13 9PL, United Kingdom*

Francis Livens, *The Dalton Nuclear Institute, The University of Manchester, Sackville Street, Manchester M13 9PL, United Kingdom*

Jon Lloyd, *School of Earth, Atmospheric and Environmental Sciences, Williamson Building, The University of Manchester, Sackville Street, Manchester M13 9PL, United Kingdom*

Katherine Morris, *School of Earth, Atmospheric and Environmental Sciences, The University of Manchester, Sackville Street, Manchester M13 9PL, United Kingdom*

Richard (Jan) Pentreath, *Plymouth Marine Laboratory, Prospect Place, The Hoe, Plymouth PL1 3DH, United Kingdom*

Joanna Renshaw, *School of Geography, Earth & Environmental Sciences, University of Birmingham, Edgbaston, Birmingham B15 2TT, United Kingdom*

Clint Sharrad, *School of Chemistry, University of Manchester, Sackville Street, Manchester M13 9PL, United Kingdom*

Jim Smith, *School of Earth & Environmental Sciences, Burnaby Building, University of Portsmouth, Burnaby Road, Portsmouth PO1 3QL, United Kingdom*

John Walls, *School of Geography, Earth & Environmental Sciences, University of Birmingham, Edgbaston, Birmingham B15 2TT, United Kingdom*

Nuclear Power Generation – Past, Present and Future[i]

JOHN WALLS

ABSTRACT

In this paper we outline the origins of the nuclear power industry in the nuclear weapons programme of the Second World War, and chart the growth of the nuclear industry through the 1950s and 1960s, and its subsequent decline during the 1970s and 1980s as a result of increasing costs and economic crisis, coupled with high profile accidents at nuclear plants at Three Mile Island and Chernobyl. We then explore the claim that we are witnessing a "nuclear renaissance", characterised by a growth in the construction of new nuclear plants in the West but particularly in Asia. Three main factors have led to arguments for nuclear energy gaining greater traction: concerns over climate change and the need to promote low carbon energy technologies; the need to enhance energy security; and the need to meet large increases in demand for electricity particularly in developing countries. We then outline six variables that have the potential to impose limits on any large scale expansion of nuclear energy. Finally we explore to what extent the March 2011 disaster at the Fukushima nuclear plant in Japan is likely to negatively impact the "nuclear renaissance".

[i]This research was funded in part by the Economic and Social Research Council under The Waste of the World programme (RES000230007). My thanks to Dr Galina Walls and Professor Roy Harrison for comments on earlier drafts.

Issues in Environmental Science and Technology, 32
Nuclear Power and the Environment
Edited by R.E. Hester and R.M. Harrison

Published by the Royal Society of Chemistry, www.rsc.org

1 Introduction

Up until a few years ago, it appeared that nuclear power no longer had a place in the energy future of the West. In the aftermath of the accident at Three Mile Island and the Chernobyl disaster, as well as the problem of significant cost overruns for new nuclear plants, and the continuing problem of nuclear waste disposal and spiralling decommissioning costs, nuclear appeared to be an industry with no viable future.[1] However in recent years we have seen the return of nuclear power as an attractive option given the urgent need to meet the increased demand for electricity, especially in developing countries, as a potential mitigation strategy against climate change and to bolster energy security. With 55 nuclear reactors currently under construction and many more ordered we frequently hear talk of a "Nuclear Renaissance".[2] Enthusiasm for new nuclear build at present is concentrated in Asia and Russia with a much slower development in Europe and North America.[ii]

In this paper, we outline the origins of the nuclear energy industry in the nuclear weapons programme of the Second World War; discuss the expansion of nuclear energy into the post war period and its role in the modernisation and industrialisation process; then chart the declining fortunes of the industry and its contemporary resurgence as a potential means of mitigating climate change. We suggest that whilst new nuclear plants will come on line in increasing numbers over the next few decades, they will be built at a much smaller pace than desired and anticipated, due to a range of factor which we explore below. Nonetheless nuclear power will continue to play a role in the energy systems of many developed and developing countries, as they try and move toward more sustainable energy systems. The extent of this role will depend on the ability of nation states to navigate the challenges that face plans for new nuclear plants.

2 Origins of Nuclear Power: The Nuclear Weapons Programme

The first nuclear reactors in what were to become the world's first nuclear powers, namely the United States, UK and the USSR, were all designed to produce plutonium for their respective nuclear weapons programmes.[iii] These initial reactors were of rudimentary design, graphite blocks into which uranium fuel was placed and plutonium chemically extracted from the spent fuel to be used in atomic bombs. The world's first nuclear reactor, built as part of the Manhattan project),[iv] achieved criticality in December 1942. Following this, a

[ii] At present there are over 440 commercial nuclear power reactors operating in 30 countries, with 376 000 MWe of total capacity. In total, they provide about 15% of the world's electricity.

[iii] The only country that developed nuclear reactors with no military link was Canada, whose ZEEP (Zero Energy Experimental Pile) formed the basis of Canada's indigenous nuclear reactor design – CANDU – which used natural as opposed to the more expensive, enriched uranium. However, the first reactor which formed part of the Manhattan projects' attempt to produce plutonium for the atomic bomb, involved scientists from Canada, Britain and France. Although Canada did not develop its own nuclear weapons programme after the war, it did sell plutonium to the UK in order to fund the Canadian civilian reactor programme.

[iv] Chicago Pile1. The term "nuclear reactor" was not used until 1952.

number of reactors were subsequently constructed at the Hanford nuclear site in Washington State, in order to produce plutonium for the first atomic bombs. The Manhattan project mobilised over 100 000 people and in today's money cost $22 billion.

In the aftermath of the Second World War, the allure of "the bomb" was strong, appearing to be the ultimate trump card reflecting a nation's prowess. Such geopolitical reasoning remains strong to the present day, witness a number of developing countries' desire to acquire nuclear weapons in order to project regional and global influence.

As a result of the research conducted during the Manhattan project, scientists in the West and the USSR realised that the heat generated from nuclear fission could be harnessed to generate electricity for power hungry nations, as well as to provide propulsion for submarines and aircraft carriers. The first nuclear reactor to produce electricity (albeit a trivial amount, enough for four light bulbs) was the small Experimental Breeder Reactor (EBR-1) in Idaho, USA, which started up in December 1951. It was, like a number of reactors in the years following the end of the war, a prototype "fast breeder reactor" designed to run on plutonium, itself extracted from spent fuel from a standard reactor. The plants were designed to produce electricity whilst "breeding" more plutonium, thus, in theory at least, they would continually produce all the fuel they needed.[v]

From the beginning, it was recognised that military and peaceful applications were intricately linked:

> *"The development of atomic energy for peaceful purposes and the development of atomic energy for bombs are in much of their course interchangeable and interdependent".*[3]

So reads a passage from a seminal report written by the then US Secretary of State, Dean Acheson in 1946, which became known as the Acheson–Lilienthal Report. It proposed transferring ownership and control of the nuclear fuel cycle from individual nation states into the hands of the United Nations Atomic Energy Commission. In principle both the USA and the USSR backed the idea, initially mooted in discussions between the allied powers during 1945. Niels Bohr, one of the leading researchers on the Manhattan project, became increasingly convinced during the war that atomic research should be shared between the USA and the USSR, primarily as a means of reconciling the two countries, even suggesting they share details of the Manhattan project be shared between the two countries.[vi]

[v] The focus on fast breeder reactors (FBR) in these early years reflected a concern that sourcing all of the uranium to power the world's nuclear reactors was going to prove extremely difficult. However, they all turned out to be too costly to operate and were beset by technological difficulties, as well as the heightened proliferation risk that would accompany a "plutonium economy". Subsequently large uranium deposits were discovered in Canada and Australia negating the original rational for FBRs.

[vi] Winston Churchill opposed the idea, suggesting to Roosevelt that he stop Bohr travelling to the Soviet Union to make his case, even suggesting at one point that he should be put under house arrest.[4]

The remarkable proposition was that the UN commission would in effect own and control the nuclear fuel cycle, from uranium mining through to reprocessing, and in effect release uranium to nations who wanted to build nuclear power plants for electricity production only. As part of this international control of nuclear technology, the US, the report suggested, should abandon its monopoly on nuclear weapons sharing knowledge with the Soviets in exchange for the Soviets not proceeding with weapons development. It seemed to be a win-win situation. Countries could take advantage of the promise of cheap base load electricity generated from nuclear power plants and the international community could nip proliferation risks in the bud.

However, the proposal taken forward in the Baruch Plan, failed.[vii] The small window of opportunity that existed for international cooperation on nuclear matters was firmly shut, ushering in the nuclear arms race and the cold war, the repercussions of which reverberate down to the present day. The US Congress in 1946 passed the McMahon Act, which firmly denied foreigners' (even wartime allies) access to US nuclear data. Individual countries had to pursue their own nuclear weapons and nuclear energy programmes with all the attendant costs and risks of "going it alone".

Wartime allies who had collaborated together on the Manhattan project began to develop their own weapons programme. For example, in the UK Clement Attlee created a cabinet sub committee, Gen 75, known informally as the "Atomic Bomb Committee" which met for the first time on 29 August, 1945. In December of that year, the committee agreed to the construction of nuclear reactors as part of the British nuclear power programme. As a result, the first nuclear reactor to come online in Western Europe, GLEEP (the Graphite Low Energy Experimental Pile) situated in Harwell, Oxfordshire, became operational in 1947 and was used for research into reactor design and operation as part of the new weapons' programme. Three years later in 1950, the "Windscale piles" in Cumbria achieved criticality. They were comprised of graphite blocks into which uranium was placed generating a chain reaction, with the spent fuel reprocessed to extract weapons grade plutonium on site, with reprocessing beginning in 1952. This enabled Operation Hurricane to take place, the first British detonation of an atomic bomb in the Monte Bello Islands on 2 October, 1952, which led to "Blue Danube", the UK's first free fall nuclear bomb, came into service in November 1953.[viii]

[vii] The plan failed for a number of reasons, including the refusal of the USSR to allow inspections on its territory, as well as the US position that it would not destroy its nuclear arsenal until it was convinced of the efficacy of international control and monitoring procedures. The talks collapsed two years after they had begun and the UN AEC abandoned. It would be a decade before a replacement body, the IAEA, was conceived and then shorn of any pretensions to global oversight of nuclear matters envisaged by Oppenheimer and Ascheson.[5]

[viii] Before the Second World War, France had invested the most money of any country in the world in the attempt to develop the first nuclear reactor, but German invasion and dispersion of its scientists meant that this honour was to become Enrico Fermi's, as part of the Manhattan Project The lack of financial resources in the immediate aftermath of the Second World War meant that French nuclear research fell well behind that of the British and Americans.

Unlike the plutonium-producing reactors in Hanford, Washington state, the Windscale piles were cooled by air being blown straight through the "piles" and discharged from tall stacks directly into the outside atmosphere.[ix] This initiated a period of massive investment in nuclear science R & D in the UK under the guidance of the Atomic Energy Research Establishment based at Harwell, which was tasked with undertaking R & D in nuclear fission for both military and civilian uses. From the late 1940s, Harwell was conducting research into reactor design for energy production.

It is clear that "without the nuclear weapons programme, and if normal commercial criteria had been applied, it is doubtful if a civil nuclear industry would ever have arisen".[6] It was realised that changing the design of these plutonium-producing piles could allow the heat to generate steam and the steam could be used to drive a turbine to produce electricity, it was these changes that formed the basis of the UK's civilian nuclear reactors.[7]

In the post war era as Britain still had to import relatively expensive oil and, to an extent coal, policy makers thought that nuclear energy could be a cheap alternative. Given its origin in the weapons' programme, the free exchange of information was curtailed. The absence of informed debate in this climate of secrecy meant that the positive aspects of nuclear energy were emphasised with negative issues rarely discussed in the public sphere. This was beneficial to governments who were keen to develop their nuclear weapons programme away from the glare of public scrutiny and to the multinational companies who, given their involvement in military applications of nuclear technology, saw profitable opportunities in new areas such as developing and selling nuclear reactors.[8,9]

If secrecy and elite decision-making surrounded the development of nuclear technology in the West, this was taken to a different level in the USSR, where a number of "closed cities" were created such as Ozyorsk (known as Chelyabinsk-65) which housed a plutonium-production plant. Soviet citizens had to have special permission to visit these cities. It was in one of such cities, Obninsk, 100 km southwest of Moscow, that the world's first nuclear power plant to generate electricity for a national grid came online. The AM-1 ("Atom Mirny" – "peaceful atom") reactor was a prototype water-cooled and graphite-moderated, with a design capacity of 30 MWt or 5 MWe. It produced just 5 megawatts of electric power. For 10 years it remained the only nuclear power plant in the USSR.[x]

During the Manhattan project, a naval officer, Hyman Rickover[xi] (later to become Admiral), realised the potential application of nuclear energy to submarine propulsion, so he initiated R & D which led to the development of what

[ix] The UK's most serious nuclear accident occurred as a result of a fire in Windscale Pile 1 in 1957. Even today, it remains a decommissioning headache in both financial and technological terms. The fire received very little media and public attention at the time, reflecting the tight security and secrecy that enveloped the nuclear industry in these early years of IES development. Unlike today, reactor designs were not subject to public scrutiny and/or parliamentary oversight.

[x] This was a forerunner of the RBMK reactors, the same design as the reactors at Chernobyl.

[xi] Rickover became known as the "Father of the Nuclear Navy".

was to become known as the pressurised water reactor (PWR) used to power the first nuclear submarine *USS Nautilus*. The PWR used enriched uranium oxide fuel and was moderated and cooled by ordinary (light) water. *USS Nautilus* was launched in 1954, three years ahead of the first commercial nuclear power station, which was also overseen by Hyman Rickover. These compact reactors which used uranium as fuel and pressurised water as both coolant and moderator evolved into the Pressurised Water Reactor (PWR) which dominated the American and other international markets. The PWR and the Boiling Water Reactor (BWR) – known collectively as "Light Water Reactors" – dominated the US and international market in reactor design and still do so today. However, a number of observers have concluded that LWRs, rather than necessarily being the best reactor design chosen after careful consideration of alternatives, were rushed forward after the concern that was generated by the first Soviet atomic bomb test.

The United States Atomic Energy Commission (AEC), created in January 1947, effectively transferred control over nuclear energy from the military to civilian institutions. Whilst in its early years the AEC's main job was to produce nuclear warheads for the military, now it was also tasked with developing and regulating civilian nuclear power, which created a conflict of interest.[xii]

This focus on military applications changed in 1953 when President Eisenhower proposed his "Atoms for Peace" programme, which reoriented research effort towards electricity generation and set the course for civil nuclear energy development.[10] Eisenhower suggested nuclear materials be used to provide "abundant electrical energy in the power-starved areas of the world". This set in train a number of international efforts at pushing this vision forward, from the Geneva Conference on the Peaceful Uses of the Atom in 1955 to the formation of the International Atomic Energy Authority (IAEA), whose mandate was to "accelerate and enlarge the contribution of atomic energy to peace, health, and prosperity throughout the world".

The optimism and almost euphoria about the possible manifold peaceful uses of the atom captured the imagination of writers and scientists, with claims that we would, aside from benefiting from cheap electricity, see "nuclear powered planes, ships, trains... nuclear energy would genetically modify crops and preserve grains and fish".[11] This "nuclear utopianism" was rarely challenged, receiving widespread support from the public and policymakers.

The cold war enabled nuclear power to be constructed as vital for national security, and the political climate generated by McCarthyism during the 1950s in the US meant that research into potential safety problems and hazards from nuclear power were discouraged.[12] Legitimate concerns over the effects of atomic testing were seen as subversive and un-American.[13] The Atoms for Peace programme was in part designed to dissuade foreign states from developing nuclear weapons. To this end the US government supplied highly enriched uranium

[xii] This critique led to the separation of regulatory and promotional functions spelt out under the *Energy Reorganization Act of 1974*, moving regulatory functions into the Nuclear Regulatory Commission (NRC) and promotional activities into the US Department of Energy.

(HEU) to countries who promised not to construct atomic bombs.[14] Suffice to say that not all of the HEU is accounted for today. This new atomic age of abundance and prosperity was also an opportunity for business to take advantage of the commercialization of the atom. The US Atomic Act (1946) was modified in 1954 to allow private sector firms to build and operate nuclear plants.[15,xiii]

3 Expansion of Nuclear Power

The large scale use of nuclear power during the 1950s and 1960s was concentrated in the USA, UK, Russia and Canada. Whilst some Western European countries began developing research programmes (often with experimental reactors), many full scale nuclear plants did not start producing electricity until the later 1960s and 1970s (Sweden, Japan, West Germany). West Germany was the first non-weapons nation to start up a nuclear power station in November 1960. However, a commitment to nuclear power is at the heart of the European Union. The Euratom Treaty signed in 1957 is one of the founding treaties of the European Union. The treaty recognised the need for an expansion in the supply of electricity for European economic growth, stating that "nuclear energy represents an essential resource for the development and invigoration of industry". It was also touted as a solution to the urban pollution caused primarily by coal-fired power stations located close to urban areas that plagued many of Europe's cities in the immediate post war period.[17]

Whist the state took control of planning and construction of nuclear plants in many European countries, in the United States the federal government was keen for the private sector to invest in nuclear power. This proved extremely difficult to do in a situation where cheap oil and coal were available to the energy utilities. As a result, the federal government financed and built a number of demonstration reactors to prove to the private sector that nuclear was feasible. As a result, Westinghouse designed the first fully commercial PWR, a 250-MWe reactor at Yankee Rowe, which started up in 1960. Meanwhile a 250-MWe Boiling Water Reactor (BWR) was designed by General Electric, Dresden-1, which also started up in 1960.[xiv]

In an attempt to build up the market for nuclear reactors the two major companies, Westinghouse and General Electric initially sustained losses of up to $1 billion per plant. This high risk strategy eventually paid off as a rush of orders from Energy Utilities ensued, amounting to 44 reactors during 1966–1967 alone (Scurlock 2007b). By the end of the 1960s, orders were being placed for PWR and BWR reactor units of more than 1000 MWe.[xv]

[xiii] The pressure on reactor designers to keep the costs down, it is claimed, led to compromises on safety, especially given the intense competition from coal and oil-fired power stations.[16]

[xiv] A prototype BWR, Vallecitos, ran from 1957 to 1963.

[xv] Not all reactors were light water reactors. Canadian reactor development headed down a quite different track, using natural uranium and heavy water as both a moderator and coolant. The first of these "CANDU" units started up in 1962 and they were the first reactors to not have a military connection. Along with Canada, Germany and Sweden followed this heavy water/natural uranium route given their desire not to have to rely on foreign states for costly enrichment services.

A pamphlet published by the nuclear company Westinghouse in 1967 captures the prevailing optimism about the promise of nuclear power at the time:

> *"It will give us all the power we need and more. That's what it's all about. Power seemingly without end. Power to do everything that man is destined to do. We have found what may be called perpetual youth".*[18]

The first full-scale civilian reactor to provide electricity to a national grid, a gas graphite reactor, came online in the UK at Calder Hall (on the same site as the Windscale plutonium piles) officially opened with much fanfare by Queen Elizabeth II on 17 October, 1956. Calder Hall eventually comprised four "Magnox" reactors (so called because of the alloy cladding around the fuel rods) which generated 50 MWe of power each, with the plants having both a commercial and military use.[xvi]

In February 1955, a White Paper *A Programme of Nuclear Power* took the engineering industry by surprise when it announced a major programme of 12 Magnox stations to be built between 1957 and 1962 (ref. 19, 20). The White Paper justified this by arguing that there would be a growth in demand for electricity that the coal industry could not meet, and that over time electricity from nuclear stations would be cheaper than coal.[21] There was a naïve assumption that nuclear plants would be no more challenging to build than coal-fired stations. The 1955 White Paper was brimming with such confidence, suggesting that the Magnox programme could contribute 25% of the nation's electricity at a cost today of $5.7 billion.[xvii] The Suez crisis created concerns over energy independence which led to increased calls for more nuclear power.[23] However, the Magnox programme was characterised by escalation in costs and time overruns which reflected problems in the tendering process, with competing consortia winning contracts at individual plants which meant novel design changes at each site, and as a result economies of scale could not then be realised.[xviii]

Sir John Cockcroft, former head of Harwell, advised Government that electricity generated from nuclear would in all probability be more expensive than alternatives (such as coal). Eventually the Labour government conceded that coal-fired power plants were 25% cheaper than nuclear. Nuclear power stations were seen as additionally useful to reduce the bargaining power of the coal miners' unions, with uranium being seen as "strike proof" given that only small amounts were needed to power reactors.[24] Indeed one of the earliest campaigns against nuclear power in the UK was initiated by the National Coal Board (NCB),[xix] which tried to expose the subsidies provided to the nuclear

---

[xvi] It was not until 1995 that plutonium production ceased.

[xvii] Eventually it estimated that a programme of twelve nuclear power stations with a total capacity of between 1400–1800 MW would be on line by 1965 (ref. 22).

[xviii] There were technical problems during operation, much longer construction times than planned and as a result a much greater cost of electricity than budgeted.

[xix] Led by the then chief economist of the NCB, Fritz Schumacher, who went on to penn the environmentalist classic *Small is Beautiful*.

industry by government. The NCB did not believe the CEGB's claim that nuclear was cheaper than coal, but were frustrated given they could not gain access to the data on costs which were covered by the secrecy laws that made it impossible to gain information on nuclear matters.

As a result of the 1964 White Paper *The Second Nuclear Power Programme*, the government chose the 600-MWe Advanced Gas Reactor, which would complement the 4190 MWe generated by the Magnox stations with 8380 MWe of AGR capacity. AGRs were eventually built at seven sites across the UK.[25] Each station was built by a different consortia which drove up costs and, as with the Magnox build, this hindered economies of scale. The White Paper on fuel policy in 1967 further reinforced the preference for nuclear over coal.[xx] A government statement to the House of Commons in 1963 stated that nuclear generation was more than twice as expensive as coal. The "plutonium credit" which assigned a value to the plutonium produced was used, initially secretly, to improve the economic case, although the operators of the power stations were never paid this credit. During this period of British history, even conservationists expressed a preference for nuclear power over coal mining in relation to the perceived lesser negative impact on the natural environment.[27] Indeed as late as 1972, an article appeared in the journal, *Environment*, arguing that "there has been very little public opposition to nuclear power in England".[xxi]

The role of government regulation and liability guarantees were central to the success of nuclear power. In the USA for example, the Price Anderson Act of 1957 which established a ceiling of $560 million for private sector liability for nuclear accidents, enabled private sector involvement in nuclear power production to proceed.[xxii]

One event was to provide a huge boost to the fortunes of the nuclear industry: the OPEC oil crises of 1973–1974. The oil crises of 1973–1974 which saw oil prices quadruple overnight made energy independence and energy security key policy issues worldwide. In France for example the result was a government review culminating in the "Messmer plan" whose aim was to secure energy independence. As a result 56 nuclear reactors were eventually built.[29]

[xx] The Prime minister at the time, Harold Wilson, and his Energy Secretary, Tony Benn, were both pro nuclear. Benn was convinced of the case for civil nuclear power based on the "beating swords into ploughshares" sentiment. A position Benn has since retracted, arguing that he was misled when Minister of Technology about the costs of nuclear energy, "I was told, believed and argued publicly that civil nuclear power was cheap, safe and peaceful and it was only later that I learned that this was all untrue since, if the full cost of development and the cost of storing long-term nuclear waste is included in the calculations nuclear power is three times the cost of coal when the pits were being closed on economic grounds".[26]

[xxi] The regulatory structure was also more permissive with regard to nuclear power than exists today, reflecting in part the deferential culture toward experts and scientists in the 1950s. This scrutiny as it existed was carried out by the UK Atomic Energy Authority internal safety branch. It relied in essence on a "staged operating experience to demonstrate that if the reactor worked, then it must be safe after all" which is in stark contrast to the risk-based approach adopted by contemporary regulators.[28]

[xxii] This was increased to $7 billion in 1988.

Prior to 1973 much of the country's electricity came from oil. The oil crisis exposed a worrying dependence on foreign states. Lacking domestic fossil fuel, they found themselves highly vulnerable to sudden spikes in the price of oil. Government leaders saw a centrally planned nuclear programme, launched through collaboration with Électricité de France (EdF), as a rational solution. The nuclear option was also portrayed as a resurrection of *la patrie*—frequent comparisons were made between reactor sites and such hallowed monuments as Notre Dame and the Eiffel Tower. During these years of expansion nuclear energy seemed to embody modernisation, industrialisation and aspirations for technological achievement. In France nuclear power was seen to contribute to the "radiance of France" to counteract the country's rapidly declining influence in world politics.[30,31] As Hecht observes, "The image of a radiant and glorious France appeared repeatedly in the discourse of engineers, administrators, labour militants, journalists, and local elected officials. These men actively cultivated the notion that national radiance would emanate from technological prowess".[32,xxiii]

As a result the percentage of electricity generated by nuclear power in France rose from 7% in 1973 to 78% by 1994 (ref. 33) a level which it has maintained to this day. Other countries such as Japan vowed to intensify their commissioning of new nuclear power stations as a result of the oil crisis, this was particularly acute in nations that had no substantial indigenous energy supplies and that were dependent on imports of oil, coal and gas. Nuclear energy has been a national strategic priority in Japan since 1973. Indeed, across Asia during the 1970s and 1980s a number of countries began to buy and licence western nuclear technologies, leading to a situation today where Japan, South Korea, India and China have burgeoning domestic nuclear R & D capabilities.[xxiv]

The period following the oil crisis then witnessed the biggest increase in nuclear plant orders even seen in France, Belgium, Sweden, Japan and the USSR.[34] In this period of exponential growth a total of 423 nuclear reactors were built from 1966 to 1985 (IAEA 2008). The USSR began selling reactors to Bulgaria, Czechoslovakia, East Germany and Poland, even to Finland.

During this period, the nuclear industry was at pains to demonstrate the benefits nuclear power had for consumers. In a 1975 survey of the 24 American utilities which operated nuclear power plants, the industry claimed that $750 million had been saved in customers' utility bills in 1974, compared with the cost had the electricity come from fossil fuels only.[35] We can see the confidence that policy makers had during this time in the ability of nuclear energy to be the main source of electricity. In 1974, the Nixon administration launched "Project Independence" which optimistically called for nuclear power to provide 50% of the nation's energy needs by the year 2000.[36] However the economic recession that followed in the wake of the Oil Crisis led to a drastic

[xxiii] As Laurent Striker, senior vice president at Électricité de France, commented "France chose nuclear because we have no oil, gas or coal resources".

[xxiv] Indeed, in December 2009 the United Arab Emirates accepted a bid from a South Korean consortium to construct four APR1400 reactors by 2020. China will reportedly invest $175 billion over the next ten years on developing the 130 square-kilometre Haiyan "Nuclear City".

reduction in electricity demand in many countries, which steadily reduced the attractiveness of nuclear plants (and for coal plants) for utilities and governments.[xxv]

Whilst we can view this period as one characterised by optimistic expansion and public acceptance, it was also a period when a "counter-expertise" was slowly forming, primarily around environmental NGOs and academics. Decision-making itself remained firmly in the hands of engineers, and governments sought to reassure the public rather than foster transparency, but in some countries there were attempts to move away from the DAD approach to decision making (decide, announce, defend). Books with titles such as *Man and the Atom: Building a New World through Nuclear Energy*, by the celebrated scientist Glenn Seaborg (1971), celebrated the progressive potential of atomic research. Other books also were published, however, which countered this optimism, such as Curtis and Hogans' (1969) *The Perils of the Peaceful Atom: the Myth of Safe Nuclear Plants*. Although the optimistic vision remained dominant, voices were being raised which had begun to question the unbridled optimism of an earlier age.

Contrary to conventional wisdom, there was even opposition to nuclear power in France, particularly at the local level. In 1976, 55% of the French population were hostile to nuclear power and the antinuclear movement had managed to penetrate local representative politics.[38] However, a number of factors, including: pro-nuclear trade unions, cross-party consensus on the orientation of energy policy and the fact that the French electoral system made it difficult for smaller parties to enter parliament, impeding this movement in gaining influence on national policy.[39]

In the UK, the secrecy and closed decision-making structure that had been a feature of the nuclear industry also began to come under challenge during the 1970s, with the government pressured to adopt a more open policy style, epitomised by the six-month Windscale Public Inquiry in 1977. The inquiry was held in response to British Nuclear Fuels' (BNFL) application to build a thermal oxide reprocessing plant (THORP) for national and international spent nuclear fuels (Hall 1986), a project which critics argued would turn the UK into the "World's Nuclear Wastebin".[40] While the inquiry was hailed by some as a "landmark in British nuclear policy making",[41] for its broad scope and participatory approach,[42] the final report was criticised for failing to justify why the arguments of the opposition had been rejected.[43]

In countries which are in the vanguard of new nuclear build today, the oil crisis did not impact upon them in the same way as it did in the West. In China for example, given its reliance on domestic coal for the vast majority of its energy needs, there was no impulse to invest in nuclear. Whilst in 1970 the then Chinese

[xxv] During this period of expansion the uranium-based thermal reactor were seen by the nuclear industry as very much the first "primitive" form of reactor,[37] in comparison with more advanced fast breeder reactors which are designed to "breed" more plutonium than they can consume as fuel (some breeders can produce 30% more fuel than they use). India, Russia, Japan and China currently have operational fast breeder reactor programmes. The UK, France and Germany have effectively shut down theirs.

premier, Zhou Enlai, argued that China needed to explore the peaceful uses of nuclear energy, the first nuclear plant did not begin construction until 1985 and only became operational in 1991.[xxvi] A number of factors can help to explain this retarded development relative to the West. State funds available to invest in nuclear plants were not made available by the Chinese government during the 1970s through to the 1990s. Chinese policymakers thought that domestic coal reserves were sufficient to meet the growing energy needs of the country, furthermore until 2005 nuclear energy was not part of the nation's strategic energy plan, prior developments being "haphazard and lacking strategic vision".[45] However, whilst public opinion in the West has waxed and waned in relation to support of nuclear power, in China public opinion is strongly supportive.[46]

In the first two decades of operation there had been reactor accidents in Canada, the UK, the USA and Switzerland, leading to a small number of deaths and millions of dollars of damage, none, however, had been of sufficient severity to throw the industry into turmoil and lead to a reversal in public confidence. This was about to change.

4 A Period of Decline

By the time of the oil crisis in 1973, seventeen countries had 167 power reactors with a capacity of nearly 61 000 MWe.[47] The nuclear industry thought the fourfold increase in oil would make nuclear more economic than coal. For a time this was indeed the case. However, the worsening inflationary environment led inexorably to a steady raising of interest rates making capital-intensive projects like nuclear stations very costly. Moreover, the deep global recession that began to bite as the 1970s drew to a close led to a steep fall in electricity demand. This negatively impacted on some countries' nuclear ambitions, with some countries like France willing and able to bear the increased costs more than others. Although there were substantial orders placed in the US for new nuclear stations during 1973 and 1974, many were never completed.[48] Indeed, no new stations were ordered in America after 1978, with many plants cancelled when 90% complete.[49] This was a rapid reversal in fortunes from the golden age, announced only in 1973, when the US AEC predicted that by the year 2000 there would be 1000 operational nuclear reactors in the US alone.

Similarly in the UK the Central Electricity Generating Board (CEGB), partly in response to the oil crisis, drew up plans during 1973 to build 32 PWRs (only one – Sizewell B – was ever built) and during 1974 The Department of Energy ordered six Steam Generating Heavy Water Reactors, a programme which was mothballed in 1978 as recession deepened with £145 million having already been sunk into the project.[50]

During the late 1970s protests against the construction of nuclear plants steadily began attracting increasing numbers of people particularly in

[xxvi] Policy changes meant that there was a shift from "military uses first" to "combining military and civilian uses", this led to a state Ministry being reorganized and renamed to become China National Nuclear Corporation (CNNC) in 1989 (ref. 44).

continental Europe – from the 60 000 who demonstrated against the construction of a fast breeder reactor in Malville, France, in July 1977 where one protester was killed, to the large scale protests across West Germany in which for the first time the police began deploying water cannons and tear gas to disperse protesters. These gathering protests were labelled by one writer at the time as the "anti nuke explosion".[51]

In the late 1970s, several countries also decided to abandon the more ambitious parts of their nuclear programmes, particularly fast breeder and reprocessing technology. This development was led by the United States when President Carter issued a decree in 1977 halting reprocessing, formally motivated by nuclear proliferation concerns but also responding to the enormous technical difficulties and financial costs that had arisen with regard to developing fast breeder reactors and reprocessing plants.[52]

From the late 1970s to 2002 the nuclear power industry suffered relative decline, particularly in Europe and North America. The number of reactors coming online from the mid 1980s little more than matched retirements, though capacity increased by nearly one third and output increased 60% due to capacity plus improved load factors in existing reactors. The share of nuclear in the world electricity market from the mid 1980s was fairly constant at 16–17%. During the period 1986–2005 there were only 71 new nuclear plants constructed, with only a small number of these being in the "old" nuclear power countries; in the preceding 20 year period there had been 436 new nuclear plants. As a result, but also due to an increase in secondary supplies, the uranium price dropped.

Moreover, in the USA, in what had been one of the largest markets for nuclear power stations, nuclear reactors became increasingly uncompetitive in comparison to coal-fired power stations as well as to the emerging combined cycle gas turbine plants.[53] Gradually "the optimism of the early 1970s turned to pessimism about the future of nuclear power".[54] Although at first the US Clean Air Act increased the attractiveness of nuclear power in comparison to coal through the 1970s, this was more often than not offset by the effects of capital cost escalation and increasing time delays that plagued the construction of new nuclear plants. From the mid 1970s to the mid 1980s 100 nuclear plants were cancelled in the US alone.[55]

There were two accidents, however, which sounded the death knell for the nuclear industry, reinforcing the negative public and utility perception toward nuclear power that had emerged during the 1970s. Firstly, the partial core meltdown in 1979 at the reactor at Three Mile Island, Pennsylvania, which, although did not lead to any loss of life, caused over 100 000 people to flee their homes and cost a great deal of public money ($2.5 billion). This hastened moratoria on building new nuclear stations in a number of European countries (Italy, Belgium and Sweden) as well as contributing to killing off the industry in the USA.

Secondly, the fire and core melt down at one of the RBMLK reactors at Chernobyl during 1986 which sent radioactivity into the atmosphere contaminating dozens of countries, causing 50 immediate casualties and thousands of cancer victims in the years afterwards, and displacing 100 000s of people across the former Soviet Union from the Ukraine and Belarus. The accident led to an intensification of anti nuclear sentiment across Europe, in Germany in particular,[56] as well as contributing to Finland (a relatively pro nuclear country) to shelve plans for a new nuclear reactor. Opinion polls throughout Europe captured the impact Chernobyl had on public attitudes toward nuclear power. Even in Finland, a country with relatively higher levels of public support for nuclear power, the amount of people who wanted to phase out nuclear power after Chernobyl increased from 21.3% in 1983 (ref. 58) to 34.5% in 1986 (ref. 59). In the UK a *de facto* moratorium on new nuclear builds was declared in 1989, pending a five year review.[59,60] Chernobyl marked the end to the ailing nuclear industry's hopes of recovery after the decline of the late 1970s. Nuclear power became increasingly framed in terms of its risks rather than its feel this is a good description benefits. The "new" framing around the risks of nuclear energy did not erase, but came to compete against and co-evolve with the original framing around the "promise of civil nuclear energy".[61]

The economic case against nuclear power became increasingly central to the debate during the 1980s. Security of supply faded to the background in a number of countries, as the influence of OPEC declined, oil prices fell, and the North Sea oil and gas fields provided cheap domestic energy. Moreover, the privatization of electricity networks also exposed the fragility of nuclear economics. The hidden costs of nuclear were starkly exposed when nuclear power stations were exempt from the privatization of the UK electricity industry in 1989, given the refusal of the private sector to take on the risk of ageing nuclear stations with potentially massive liabilities.[xxvii] In the UK, as a result of the 1995 white paper *The Prospects for Nuclear Power in the United Kingdom*, seven AGR stations and the one PWR were floated on the stock market as *British Energy* for the sum of just £2.1 billion in 1996 (Sizewell B alone had cost £2.8 billion). The private sector had got eight stations for the price of one,[65] reflecting the desire of government to get rid of the nuclear plants as quickly as possible.[xxviii] It became clear that private utilities found it more economical to build gas-fired plants than either nuclear or coal-fired plants. It also became transparent that, bereft of a subsidy, the utilities in liberalised energy markets did not view nuclear stations as the most cost effective investment.[xxix]

[xxvii] Some industry insiders suggest that the problem was exacerbated by the introduction post-privatization of a non-technical management, whose focus was on short-term profits, "with no understanding of the need to technically maintain the assets and the skill base or the long-term needs of the business, which in turn led to the massive shareholder losses".[62]

[xxviii] The ageing Magnox stations with less than 10 years' lifespan could not be sold and were given to BNFL.

[xxix] Exacerbated by the lack of a long-term waste disposal route and a regulatory regime that hindered rather than facilitate new nuclear plants.

5 A Nuclear Renaissance? Possibilities and Challenges

In the 21st century several factors have combined to revive the prospects for nuclear power. First is the realisation of the scale of projected increased demand for electricity worldwide, but particularly in rapidly developing countries, for which nuclear is increasingly seen as part of the solution. Secondly, there is a raised awareness of the importance of energy security and thirdly, the urgent need to encourage low carbon energy generation technologies to mitigate the threat of dangerous climate change. Over the last decade the increasing arguments for nuclear and the pace of new orders for nuclear reactors has led talk of a "nuclear renaissance".[64] At the turn of the millennium the nuclear industry and its supporters recognised that a more proactive approach was necessary to ensure public acceptance, as well as the need to promote nuclear as an attractive option for policymakers. Books such as *Preparing the Ground for Renewal of Nuclear Power*[65] reflected the growing confidence of the nuclear industry.[66] A confidence which increased with the public support of prominent environmentalists, "nuclear power is the only green solution",[67,xxx] and, moreover, that "the worst possible nuclear disasters are not as bad as the worst climate change disasters".[68] A raft of government sponsored studies such as *Options for a Low Carbon Future* reinforced the march of nuclear, viewing nuclear as a potential weapon in the struggle to reduce carbon emissions in the battle against climate change. One prominent report argued that:

> *"The present study has confirmed the Royal Commission on Environmental Pollution (RCEP) conclusion that the replacement of the current nuclear power stations by new nuclear stations and an expansion of nuclear power could help the UK reduce its CO2 emissions by 60% or more by 2050".*[69]

The attractiveness of nuclear power then, as the Intergovernmental Panel on Climate Change (IPCC) suggests, is because "the life cycle GHG emissions per kWh from nuclear power plants are two orders of magnitude lower than fossil-fuelled electricity generation and comparable to most renewables".[70]

If we are to shift from a reliance on coal and gas which are the dominant sources of electricity across the globe, what are we to use? Demand reduction in the form of energy conservation can play an important role here but we are still left with replacing supplies from fossil fuels.[xxxi]

France is often held up as an example of what can be achieved in terms of emissions reductions when nuclear power forms a large part of the

---

[xxx] However, even a doubling of existing nuclear capacity will only reduce GHG emissions by 8% given that electricity is only a third of total energy production.

[xxxi] Whilst it is clear that nuclear power is not a completely carbon-free energy source (*e.g.* both uranium mining and the construction of the nuclear plant relies on fossil fuel energy), it is substantially better than either coal or gas.

electricity supply, for it generates 75% of its electricity from nuclear power and emits 6.6 tonnes of CO2 per capita, compared with 10.4 tonnes per capita for Germany.[71] A recent report from the MIT on the *Future of Nuclear* suggests that nuclear-generating capacity be increased almost three-fold, to 1000 billion Watts by 2050, thereby avoiding 1.8 billion tonnes of carbon emissions annually from coal plants (about 25% of the increment in carbon emissions otherwise expected in a business-as-usual scenario).[72,xxxii] From a strict climate change perspective, nuclear power is an improvement over conventional coal-burning power plants. A nuclear power plant does not directly produce greenhouse gas emissions (unless it is running idle, being refuelled or operating on backup generators) and it emits about one-tenth to one-twentieth the carbon dioxide emissions over the course of its lifecycle as compared with a comparatively sized conventional, fossil-fuelled power plant.[74,75] Still, reprocessing and enriching uranium requires a substantial amount of electricity, often generated from fossil fuel-fired power plants, and uranium milling, mining, plant construction and decommissioning all generate greenhouse gas. A recent review which assessed the most cost effective low carbon base load electricity-generating technology concluded that "nuclear energy is the cheapest option and best able to meet the IPCC timetable for GHG abatement".[76] Whilst there would be large financial costs involved in any new nuclear programme, proponents argue that all of these potential costs are insignificant compared with the risks posed by climate change.[77]

As a result of these debates we have witnessed a change in the nature of the public discourse around nuclear power over the past decade. Those factors which previously led people to reject nuclear power (such as costs, waste disposal, accidents, proliferation concerns, *etc.*) are being discounted because of the way in which nuclear power has been reframed and repackaged as a solution to climate change, energy security[78] and as a means of meeting the increasing demand for electricity in developing countries.[79] All of which is reflected in the more positive public attitudes toward nuclear power that are shown in opinion polls.

Surveys conducted in the UK in 2005 (ref. 80) and 2006 (YouGov), found that 35% and 40% of the respondents, respectively, were in favour of new nuclear. This rose to 62% and 68% if nuclear new build was coupled with a concerted policy of promoting renewables.[81] The latest opinion poll in the UK shows the highest level of public support for nuclear power in over a decade, with 40% of people favourable to nuclear (up seven points from 2009), while 17% are unfavourable (down three points).[82] In 2008, 43% of Finns supported new nuclear build, while 25% wanted to phase out

[xxxii] In some countries (such as Finland), potential increases in natural gas prices played a key role in the decision to proceed with new nuclear. In addition, nuclear power was portrayed as the cheapest low carbon option.[73] In May 2002, the Finnish parliament voted on the new reactor and decided in favour, becoming the first OECD country to decide to build a new nuclear reactor for several years. The vote was very close, however, with 107 votes for and 92 against.

nuclear.[83] Public support as monitored by the Eurobarometer polls suggests that in 8 out of 25 countries of the EU there is a majority in favour of nuclear power.[xxxiii] However this increase in optimism and support for nuclear power has been negatively impacted by the Disaster that occurred at the Fukushima nuclear power plant in Japan during March 2011. (As the author received proofs for this paper the devastating Tsunami hit Japan on March 11th 2011, thereby setting in train a sequence of events that led to the catastrophic loss of coolant in a number of nuclear reactors at the Fukushima Daichi nuclear plant, on the west coast of Japan, leading to contamination of the site and surrounding area. The disaster has now been classified as a 7 on the severity scale for nuclear accidents, the same as Chernobyl.) Whilst the medium and longer term impacts on public perceptions of nuclear energy are unknown, the immediate response has been stark. In the week following the disaster a Gallup poll found that "support for nuclear energy worldwide has fallen from 57 percent before Japan's nuclear disaster to 49 percent now, but supporters still outnumber opponents". However there are large regional and country differences. In America for example a recent poll found that 44% of the public were in favour and 47% opposed to "the construction of nuclear power plants in the United States",[167] this is down from a high of 62% from just a year earlier.[168]

Within the EU a recent poll found that whilst the Fukushima nuclear disaster has led to greater worry across the European Union about the safety of nuclear power plants, with the exception of Germany – citizens were broadly confident about the management of nuclear plants in their own country. Indeed, "66% of Germans now oppose nuclear power, while just 19 per cent support it. The French are evenly split on the matter, at 36 per cent each, while the British are in favour by a margin of 35 per cent to 30 per cent".[169]

Worldwide there are 60 new nuclear plants under construction with 131 more proposed, a number of which are in countries which do not have nuclear power *e.g.* United Arab Emirates, Bangladesh, Vietnam, Egypt, Indonesia, Thailand and Turkey (Montgomery 2010).[xxxiv] The global recession has dented the hopes

[xxxiii] Bickerstaff *et al.* describe the British public attitude toward new nuclear as "reluctant acceptance", *i.e.* when presented against the impending danger of climate change, the risks of nuclear power seem acceptable even to people who are *a priori* hostile to nuclear power. While science itself is trusted, the government institutions are seen as "unreliable, secretive and failing to execute their proper duties (or functions) to serve the public interest".[84]

[xxxiv] Whilst a recent report from the International Atomic Energy Agency (IAEA) *International Status and Prospects of Nuclear Power* suggested that there are some 65 countries currently without nuclear power plants who "are expressing interest in, considering, or actively planning for nuclear power" there are technical barriers to at least 17 of those proceeding, due to the fact that they have electricity grids of less than 5 GW which are "too small to accommodate most of the reactor designs on offer". Moreover, many of these countries do not have the "necessary nuclear regulations, regulators, maintenance capacity, or the skilled workforce to run a nuclear plant. The head of France's Nuclear Safety Authority has estimated that it would take at least 15 years to build the necessary regulatory framework in countries that are starting from scratch".[85] There are also doubts as to whether grids of up to 10 GW could cope with nuclear power generation.

of some countries who have decided not to press ahead with further nuclear expansion after initial investment, *e.g.* Turkey.[xxxv] The new build programme in Europe (excluding Russia)[xxxvi] amounts to just six reactors in four countries: Finland, France, Romania and Slovakia. The new build programme in Europe began in 2004 when the first of the late third-generation units was ordered for Finland – a 1600 MWe European PWR (EPR). A similar unit is being built in Flamenville, France, with another on order as part of the replacement for the PWRs built in France during the 1970s and 1980s. There are plans for a new build programme in a number of countries such as the UK, Bulgaria, Czech Republic and Slovenia. Prior to the Fukushima disaster there were countries (such as Italy and Sweden)[xxxvii] who were considering the revival of mothballed nuclear programmes. Even in what has become a staunchly anti nuclear country such as Germany, who only a decade ago pledged a "comprehensive and irreversible" end to nuclear power, had pledged to extend the life of its existing reactors by an average of 12 years. Nuclear power is to act as a bridge in order to allow renewable energy to eventually provide most of the country's energy needs by 2050 (up from 16% today). From this perspective, nuclear buys time for renewable energy to fully develop and mature, "while slowing down the worst effects of global warming".[86]

However as a result of the Fukushima disaster a number of countries have imposed some limits to prolonging the operating life of existing nuclear plants and/or placing limits on future new build proposals. Germany for example, one of the most anti-nuclear countries in Europe, has imposed a 3-month moratorium on the reactor lifespan extension passed in 2010. Moreover, Germany has also decided to temporarily shut down 7 of its 17 reactors, with Italy imposing a one year moratorium on the construction of new nuclear power plants. A number of other countries have tasked their regulatory authorities to investigate the disaster in order to learn lessons that can be applied to increasing the safety of domestic nuclear plants (UK, USA and France) with a small number content to proceed with new build proposals such as Slovakia with China announcing a pared back nuclear expansion programme. A recent report issued by UBS suggests that at the very least around 30 nuclear plants may have to close as a result of Fukushima, in particular those in seismic zones or close to national boundaries.[170]

Plans in Europe and North America are overshadowed, however, by those in China, India, Japan and South Korea. The centre of gravity in building new

[xxxv] For many developing countries the language of national economic development is often invoked as a rationale for investing in nuclear energy, even in countries with no history of the technology.

[xxxvi] Russia's neighbour, Ukraine, is currently building two reactors and planning as many as 11 more by 2030 as it seeks to reduce its dependence on energy from Russia, particularly in light of the disputes over gas in 2006 and 2009. The strategy also envisages completing the construction by 2017 of two reactors at Khmelnitsky, work on which has been halted since 1990.

[xxxvii] A recent MIT report on the *Future of Nuclear Power* pointed out that given increasing demand, to increase nuclear powers' share from its present 17% of world electricity to just 19% by 2050 would require a near-trebling of nuclear capacity: 1000–1500 large nuclear plants would have to be built worldwide.

nuclear plants has shifted toward Asia[87] and to China in particular. China alone plans a six-fold increase in nuclear power capacity by 2020, and has more than one hundred further large units proposed and backed by political determination and popular support. A large portion of these are the latest western design, expedited by modular construction. One can say that the history of nuclear power starts with science in Europe, blossoms in UK and USA with the latter's technological might, languishes for a few decades, then has a new growth spurt in East Asia. China and India, with nearly half the world's population and determined policies to increase electricity and decrease poverty, have led the way in new build. China has 17 third-generation reactors under construction with 124 planned or proposed.

Recently it has become official policy in China to transform nuclear energy development from "active" to "aggressive".[88] The economic boom has ensured that there are sufficient funds to cover the capital costs of building nuclear plants, with preferential tax policies available for the companies involved.[89] The large sums of money that the three state owned companies are looking to invest ($117 billion in the case of the Chinese National Nuclear Corporation alone) has meant that at least one of them is looking to launch an initial public offering (IPO) in order to raise funds on the international markets. The move to build new nuclear plants was inspired in part by the experiences of 2002, when "blackouts rolled in and factory lights flickered; the grid sucked dry by a decade of breakneck industrialization. Oil and natural gas were running low".[90]

China's electricity consumption quadrupled between 1980 and 2000. The cumulative impact of air pollution as a result of burning fossil fuels is estimated to kill 750 000 people a year and economic loss is put at 6% of GDP.[91] Moreover, around 30% of China is polluted by acid rain due to the large amounts of sulfur dioxide produced by coal-fired power stations,[92] with three coal-fired stations coming online each week in China.[94,xxxviii] A recent study by BP suggests that "China can only continue at current rates of production for 38 years before its coal reserves are exhausted. That compares with 245 years in the USA and 105 years in India".[95]

In order to try and mitigate the problems of energy security, environmental pollution and climate change, China is seeking to acquire security of supply from a variety of sources, recognising that coal will remain the dominant source of energy, with nuclear only supplementing this. The transport of coal is major headache as the coal reserves are mainly located in the north or northwest, with nearly half the country's rail capacity being used to transport coal.[96]

This situation contrasts with that of the USA, the leader in the first wave of nuclear energy production, where there have been 17 applications to the Nuclear Regulatory Commission (NRC) for joint construction and operating

[xxxviii] China has approximately 25 000 coal mines employing 3.4 million people. Observers have suggested of the decline in availability of coal after 2020 that "no fossil fuel other than coal will be able to provide sufficient energy to sustain current economic growth rates in the years ahead, and non-fossil sources will require unprecedented and perhaps unachievable levels of investment just to make up for *declines* in coal production – never mind providing enough to fuel continued annual energy *growth* of seven to ten percent per year".[93]

licences for about 25 new nuclear power reactors. However, these plans seem rather ambitious given the recent announcements from key private sector partners in the new build programme that they have withdrawn due to projected overruns in project costs. Only one reactor is actually being built, at Watts Bar in Tennessee.

In fact one of the main obstacles to an ambitious new build programme in North America and Europe is the cost of financing nuclear plants. This consideration over finance has led to private sector contractors pulling out of proposals in America, to states such as Bulgaria unable to find companies willing to invest in nuclear plants at proposed rates to the reactors being built in France and Finland which are over time and cost. The response has been for governments to implement direct or indirect subsidies to facilitate investment which has drawn criticism from civil society groups and other energy providers. There are also concerns over the price and availability of uranium to power nuclear plants, and in shortages of skilled labour and reactor vessels. In what follows, we explore a number of the barriers to the future of the nuclear renaissance and the implications for nuclear power in the transition to a low carbon energy system.

5.1 Uranium: A Sustainable Energy Source?

Nuclear power depends upon a finite resource: uranium. Uranium is ubiquitous on the Earth. It is a metal approximately as common as tin or zinc, and it is a constituent of most rocks.[97] Central to assessing to what extent the current expansion of nuclear power is sustainable is an assessment of the given reserves of uranium. Current usage is about 68 000 tonnes of uranium per year, with current resources of uranium estimated at 5.4Mt. At current rates of consumption this will last 80 years. However, further exploration and improvements in extraction technology are likely to at least double this estimate over time, particularly as rising prices incentivise mining firms to increase exploration. Uranium exploration has occurred in cycles: initially driven by military needs from 1945 to 1958, then by the needs of civilian plants between 1974 and 1983, with next to no uranium exploration occurring between 1985 and 2003. As a result of growing global demand due to new build programmes in Asia, a new exploration cycle is underway with uranium mining companies embarked on renewed exploratory work, drawing upon new technology and new geological analyses.[98,xxxix] As a result of this increase in investment and exploration between 2005 and 2006, the world's known uranium reserves increased by 17%. Moreover:

> *"The price of a mineral commodity also directly determines the amount of known resources which are economically extractable. On the basis of analogies with other metal minerals, a doubling of price from present levels*

[xxxix] "In the third uranium exploration cycle from 2003 to the end of 2009, about US$ 5.75 billion was spent on uranium exploration and deposit delineation on over 600 projects".[100]

could be expected to create about a ten-fold increase in measured economic resources, over time, due both to increased exploration and the reclassification of resources regarding what is economically recoverable".[99]

The Australian Uranium Information Center suggests that if the price of uranium were to double we could expect to see a ten-fold increase in known resources: that is an increase from 3 to 30 million tonnes.[101] At various times the nuclear industry has used concerns about the long-term availability of fissile material to justify the development of Fast Breeder Technology. This type of reactor can be started up on plutonium derived from the spent fuel from conventional reactors and operated in closed circuit with its reprocessing plant. Such a reactor, supplied with natural or depleted uranium for its "fertile blanket", can be operated so that each tonne of ore yields 60 times more energy than in a conventional reactor. Breeder reactors could match today's nuclear output for 30 000 years. It is estimated that electricity from FBRs would cost around three times the amount per kilowatt as that from conventional nuclear power plants; a great deal of investment in R & D is necessary before commercialization is a possibility. India, along with a number of other countries, is currently undertaking extensive R & D in this area. However, this cannot be relied upon in either the near or medium term.[xl] Moreover, a recent report from MIT suggested that the rationale for breeders was based on an out-of-date understanding of uranium scarcity, given that it would take an LWR 30 years to provide the plutonium to start one such breeder reactor which had proven uneconomic.

The report suggests that a more effective and efficient plant would be an enriched uranium-initiated breeder (with a unitary conversion rate). In this design, natural or depleted uranium could be added to the reactor core at the same rate enriched uranium is burned up, and producing no excess nuclear material. This, suggests a recent report from MIT "is a much simpler and more efficient self-sustaining fuel cycle". They suggest there is plenty of uranium to sustain even the most optimistic worldwide nuclear power scenarios "for much of this century at least".[103]

There are additional technical solutions. For example, using more enrichment work could reduce the uranium needs of LWRs by as much as 30 percent per tonne of low enriched uranium (LEU). And separating plutonium and uranium from spent LEU and using them to make fresh fuel could reduce requirements by another 30 percent. Taking both steps would cut the uranium requirements of an LWR in half.

Known reserves of uranium are found in relatively stable industrialised countries (Australia 23%), Kazakhstan (15%), Russia (10%), Canada (8%), South Africa (8%) the USA (6%). As Montgomery indicates, "This distribution may not please everyone in New Delhi or Qinshan, but it does at least promise a reliable supply from stable nations".[104] Regardless of whether the countries are "friendly", the very fact that countries have to import uranium

[xl] The extraction of uranium from seawater would create 4.5 billion tonnes of uranium – a 60 000-year supply at present rates but it is currently uneconomic.[102]

deals a blow to claims that nuclear energy enables one to achieve "energy independence", particularly for countries like France where energy independence was the primary rationale for the headlong rush into new nuclear during the 1970s and 1980s.

The investment bank RBC Capital has recently said that the uranium market has moved from oversupply to undersupply in just a few months as China has begun to purchase long-term supplies for new reactors. The rise in price is also due to reports in the Chincse media that the country is going to build 60% more megawatts of nuclear power by 2020 than previously thought.[xli] Some countries face more difficulties than others in sourcing uranium, for example under a treaty with Russia, the United States currently receive 40% of its reactor fuel from decommissioned Russian nuclear weapons. The treaty runs out in 2013. Uranium mines only provide two thirds of worldwide uranium requirements with the rest coming from military sources (stores of uranium/plutonium and uranium, from decommissioned nuclear missiles).

5.2 *Nuclear Power Economics*

There is no escaping the fact that nuclear power plants are very expensive to build relative to all other forms of electricity production. Actual and projected costs vary depending upon a range of factors, but the front-loaded cost structure (high initial investment then relatively low running cost) according to a recent report from MIT assumes that a nuclear reactor costs $4000 per kilowatt of electricity produced to build – or $4 billion for a typical one-gigawatt nuclear power plant. Actual industry estimates for reactors being built today are at least $6 billion and as much as $10 billion. "If you build a nuclear power plant and operate it well, it's going to produce a steady stream of income", but "the disadvantage of nuclear is the enormous capital commitment that is made up front".[105]

As the MIT report goes on to say, "the track record for the construction costs of nuclear plants completed in the USA during the 1980s and early 1990s was poor. Actual costs were far higher than had been projected... The first few US plants will be a critical test for all parties involved".[106]

However, "new nuclear plants are designed to last longer without major upgrades – up to sixty years, *vs.* thirty to forty years for gas, coal, and most renewables. And they can generate as much or more power than the largest fossil plants, up to 1.6 GW or more, while saving 7–8 million tonnes a year of carbon emissions".[107] Due to their capital intensity, long construction times and proclivity for cost overruns, nuclear reactors are extremely expensive to build and are plagued by uncertainties for investors.[xlii]

[xli] This led RBC Capital to argue that uranium spot prices will double by 2012. China recently entered into a contract with the French firm, AREVA, which will see the French firm sell $3.5 billion worth of uranium to China.

[xlii] Nuclear also has the additional burden of having to set aside funds directly, or in some countries *via* a government imposed levy on electricity production, in order to pay for the costs of waste disposal and decommissioning of the plant.

The financial risks of new nuclear reactors has led Wall Street firms to indicate that it will be difficult to sell bonds to support these projects in the capital markets. Even in the United States, which under its Energy Policy Act of 2005[xliii] authorized $18.5 billion in loan guarantees to cover 80% of building the first tranche of new reactors, the incentives have not been enough to catalyze significant private sector interest investment thus far.[109] The market remembers the Washington Public Power Supply System fiasco, when in 1983 it defaulted on $2 billion worth of bonds as a result of mothballing four power plants after a string of delays and cost overruns. According to a major engineering organisation, UK utilities have cut back their nuclear build connection plans by 28% to 18.4 GW, with a possibility that this may further reduce to only 13 GW when final decisions are taken. Furthermore, industry requires some surety that government will remain committed to a nuclear future. Such concerns are exacerbated by "major uncertainties surrounding individual site planning, grid connection, licensing, waste and finance which need to be resolved without further delay and excessive consultation".[110,xliv]

It is clear that in a fully liberalised energy market with government setting no targets or market incentives, utilities would not build nuclear power plants. This is why carbon trading is extremely important for the future economic viability of nuclear power because it pushes up the cost of its two main rivals: coal- and gas-fired stations. Without a price being attached to carbon, *via* the European Emission Trading Scheme (ETS) or some other mechanism, coal and gas are the cheapest options with nuclear in third place. However, if carbon were to be 25 euros tonne^{-1} then nuclear becomes the cheapest. One of the measures in the UK government's reshaping of the energy market is precisely to impose a "floor" price on carbon.[112]

The projected *versus* actual costs of reactors do not make for good reading. One assessment explored estimated and actual US nuclear power plant construction costs from 1966 to 1977 (when the majority of American reactors were built) and found that in every case plants cost at least twice as much as expected.[113,114] The last reactor to be built in the UK, Sizewell B, cost £1.8 billion as opposed to a projected £300 million, and the new EPR reactors being built in Finland and France are both over time and budget. Assessing costs in

[xliii] Moreover, the 2005 Energy Act also streamlined the regulatory approvals process given the long delays in previous builds. Under the previous regime, once a plant had been constructed objections could be raised. The law also allows for a "risk insurance" to protect utilities against unexpected federal or state regulatory delays for up to six new reactors built under the new licensing structure. The loan guarantees provide 80% of the project cost to be repaid over 30 years. However, these loan guarantees are not having the desired effect, witness the recent decision by Constellation Energy Group Inc to pull out of a joint venture with EDF to build new nuclear power plants, citing costs of government loan guarantees to build the plants. This highlights the complex and at times unfavourable economics in a liberalised energy market even with government subsidies.[108]

[xliv] "The Government intimated that, consistent with its non interventionist free market approach, it would play its part as a market 'enabler' by reducing or removing any barriers impeding the new build programme".[111]

Asia and Russian is much more difficult given that they do not have liberalised energy markets.[115] However, concerns over the negative PR engendered by the delays in France and Finland have led some of the major plant construction companies to take out full page adverts in the leading nuclear industry magazines, such as *Nuclear Power International Magazine*, emphasising the claim that in China new nuclear plants are being built on time and on budget.[xlv] In the USA whilst a small number of proposals for new plants are proceeding, an equal number have collapsed due to some partners in the consortia deciding the numbers do not stack up, even with the existence of federal loan guarantees. Indeed, Florida Power & Light (FPL) has suspended work on a new two reactor project citing the "deteriorating regulatory environment".[116]

One of the challenges that corporations face when building new plants, is what the industry itself calls "first-of-a-kind engineering" (FOAKE) challenges. The areas of greatest uncertainty are the design and construction, financial backing, the political environment and severe weather during construction, because it is difficult to estimate their total overall impact to the project and to effectuate reasonable control. Capital cost of a "nth-of-a-kind" (NOAK) reactor should be 10 to 20% less than that of a first-of-a-kind (FOAK) reactor because of the lessons learnt in the construction and deployment of earlier units.[117] Moreover the "manufacturing learning curve generally flattens out after five to seven repeat units have been built...", even so nuclear reactors are expensive to build and vulnerable to fluctuations in national interest rates. As other countries have demonstrated (*e.g.* France and Japan), the cost structure of nuclear is such that only a large new build programme can generate "sufficient economies of scale to compete with gas-fired generation technology in the absence of effective carbon trading. In one sense, nuclear power is an all-or-nothing option".[118]

In Asia, current real-world costs are significantly lower than in North America and Europe. The two leading reactor designs now being built in China are the indigenous CPR-1000 and the Westinghouse AP-1000. Reported capital costs are in the range of \$1296–\$1790 kW^{-1} (ref. 119). Korea has focused attention on its APR-1400 design, with domestic overnight costs of \$2333 kW^{-1} (WNA 2010d) A recent contract for \$20.4 billion has been signed with Korean consortium KEPCO to build four APR-1400 reactors in the United Arab Emirates (UAE), at a turnkey cost of \$3643 kW^{-1}. This price is notable considering that it is offered under near-FOAK conditions, because these will be the UAE's first nuclear plants.

There is a great deal of regional variability then in the capital cost of building nuclear reactors. A significant part of the cost is wages for the large number of skilled and unskilled workers necessary to build and operate the nuclear plant. In addition, there are concerns that there are not enough skilled workers for an expanded new build programme.

[xlv] "In China, four new AP1000s are currently under construction and they are being built in an on-time and on-budget manner, with the first scheduled to come online as planned in 2013". (*Nuclear Power International Magazine*, 2010).

The financial effects of an increase in insurance for the viability of new nuclear plants in light of Fukushima are at present unknown however one recent review suggests that "it is hard to see how a regulatory body could give generic approval to any new design until the problems at Fukushima have been thoroughly understood and designers find ways to prevent a repeat of events. As a result, completion of the U.S. and U.K. generic reviews will be delayed. The extent of additional design requirements remains to be seen, but it is very likely that additional costs, perhaps significant, will be imposed on any new designs".[171]

5.3 Shortages in Skilled Labour and Materials

A global bottleneck in the supply chain could derail the current plans for new nuclear build. These include a lack of skilled engineers, as well as a backlog in orders for machine parts and for reactors vessels. From an engineering standpoint, the larger the order book for new reactors the more cost effective it is for companies to invest in developing the skills of the workforce. In the USA, a 2009 survey by the Washington-based Nuclear Energy Institute showed that 38 percent of the current workforce in the nuclear industry will be eligible for retirement by 2014. This has led to tailored two-year education and training programmes being rolled out at technical colleges across the USA.[120] However, such workers will be in demand globally; already a significant amount of American skilled nuclear engineers and managers work in the UK decommissioning sector, given that the UK ran down its indigenous nuclear skills base as a result of the stagnation of its nuclear industry.

Some countries are more susceptible to labour shortages than others, which may impact upon ambitious new build programmes. France faces an acute shortage of skilled workers with some 40 percent of EDF's operators and maintenance staff retiring by 2015. This has led France to advertise for overseas students to study for masters degrees in nuclear engineering and related subjects in France. "The need for students in atomic energy is estimated at 1200 graduated students a year for the next 10 years, although, nowadays, the number of graduated students is 300 per year".[121] If France can tempt overseas students to work in its nuclear power programme, then its potential foreign clients to whom it wants to sell its own nuclear technologies, won't have a nuclear workforce of their own![122]

In developing countries which are pursuing an aggressive nuclear expansion policy, the shortage of skilled labour is particularly severe. In China for instance, many students who enrol for nuclear engineering programme at university end up changing their majors, with only 30% staying in the field.[123] This is a potential problem given that China will need at least 6000 nuclear engineers by 2020.[124]

Gaps and supply chain problems can lead, as they have in the past, to inflation in the cost of key parts and to delays in construction. A shortage of trained personnel can drive costs further up. Today, for example, only one facility in the world (Mitsubishi Heavy Industries, Ltd) has the forging capability to manufacture large reactor vessels,[xlvi] which raises questions about the ability of the firm to meet the increasing global demand for reactor vessels. More investment is necessary in plants that are capable of manufacturing reactor vessels. As more and more existing reactors are having their operating lifetime extended there is also an increasing demand for a range of replacement nuclear components that require advanced heavy machinery manufacturers.

5.4 Nuclear Safety

Since its inception people have asked the question: is nuclear power safe? During the forty years that commercial power plants have operated worldwide, there have been about eight dozen major accidents, three of which –Windscale (1957), Chernobyl (1986) and Fukushima (2011) – were especially serious. At Chernobyl, 31 worker deaths were recorded immediately and it is estimated that there were up to 5900 more in the months and years that followed as a result of radioactivity from the fire in the reactor across Europe. Reactor designs since the accident at Three Mile Island (1979) have been upgraded with new, multilayered and redundant safety systems, developed specifically to avoid such errors. Chernobyl, for example, was built without any containment structure, according to a design that would have been instantly condemned anywhere in the West. Both incidents, however, inspired many changes, leading the IAEA to establish a global network for peer review of sites, designs, operating procedures and creating a collection of "best industry practices". Plant design now stresses passive safety; the reactor shuts down automatically in case of any irregularity.

Nonetheless such guidance proved ineffective in preventing the disaster at Fukushima, where worse case scenarios were discounted in the disaster management plans of the site and as a result sea defences were woefully inadequate. As a result, this led to back up diesel power systems failing due to the sea defences being overwhelmed with the wave from the tsunami and with the final battery power running out, a number of the reactors were left without coolant, leading to a partial meltdown of reactors on the site. As a result Japan has officially admitted the Fukushima nuclear disaster is as bad as Chernobyl and has upgraded the incident to the worst case level 7.

Whilst we have seen the development of a much stronger safety culture within the industry over the last 20 years, nuclear power plants have the disadvantage of being so complex that almost every reactor has experienced some sort of incident or failure over its history, and even if the risk of a true meltdown is low, the impact of such an accident would be very large.[125] Nonetheless the safety record of existing nuclear reactors has improved over time as their margins of error have improved and safety regulations have been upgraded. The industry has incorporated research findings on human factors and safety

[xlvi] A problem discussed by the Committee on Climate Change Science and Technology Integration 2009.

culture through groups and organizations such as the IAEA and the World Nuclear Association of Nuclear Operators created after the Chernobyl accident in 1986.[126]

A scenario envisaged by the authors of the MIT study was an optimistic three-fold increase in the world nuclear fleet capacity by 2050. They concluded, after undertaking a probabalistic risk assessment (PRA), that one would expect four core damage accidents during this time (they based their analysis on current estimates of core damage to occur once in every 10 000 reactor years). They concluded that this was an unacceptably high number – it should be 1 or less, which is the current expected safety level.[127] They concluded that a core damage frequency of 1 in 100 000 reactor years is a desirable goal, which is a ten-fold reduction from current levels. The designers of the new light water reactors currently being built argue that they already achieve these goals through advanced safety measures and greater use of passive safety mechanisms.[xlvii] It is beholden on regulators to assess these claims to enhanced safety of the latest generation of LWRs during the (pre)licensing phase of reactors.

We do not believe there is a nuclear plant design that is totally risk free. This is due to technical and workforce issues. Safe operation requires effective regulation, management who is committed to safety and a skilled work force. The restructuring of electricity sectors around the world has motivated some operators to place profits before safety. Undue solicitude for profits of the licensee has played a large role in explaining the mishaps that have occurred at nuclear power plants. Nuclear power is least safe in environments where complacency and pressure to maximize profits are the greatest. It is of continuing concern as to "whether nuclear reactor safety goals are compatible with the transition to competitive electricity markets".[129] Owners and managers of nuclear plants respond that it is economically beneficial to ensure high levels of safety, given the enormous financial costs of accidents. However, well funded regulatory agencies are vital to ensuring plant operators do not neglect safety inspections for continuous plant operation.

However at Fukushima it would appear that a number of regulatory failures and cost cutting exercises by the owner Tokyo Electric Power Company (TEPCO) contributed to the severity of the disaster. As far back as July 2000, "four ominous unexpected shutdowns occurred, some releasing unacceptable radiation levels, in the plants run by Tokyo Electric Power Company (TEPCO), Japan's largest utility. In 2001, a whistle-blower triggered disclosures of falsified tests at some of the company's seventeen plants, and the government forced TEPCO to close some plants".[172] Moreover, "in 2002, the company predicted that *all* of its seventeen plants might have to be shut down for inspection and repairs, because of falsified inspections and concealment of faults found in inspections that the government ordered; some of the faults were potentially catastrophic" (*ibid*). As a result a top

[xlvii] "Additional gains may come with the introduction of High-Temperature Gas Reactors (HTGRs). In principle the HTGR may be superior to the LWR in its ability to retain fission products in a loss-of-coolant accident, because of fuel form and because core temperatures can be kept sufficiently low due to low power density design and high heat capacity of the core, if RD&D validates this feature".[128] The HTGR also has an advantage compared to light water reactors in terms of proliferation resistance.

company official was charged with giving specific orders to hide large cracks in the "shrouds," or steel casings around the reactor core, in two of the thirteen reactors at which false inspection reports had been filed. According to documents from Tokyo Electric Power (Tepco), the company "repeatedly missed safety checks over a 10-year period up to two weeks before the 11 March disaster, and allowed uranium fuel rods to pile up inside the 40-year-old facility".[173] This exposes the problem of cost cutting initiated by the chief executive, Masataka Shimizu, in that the company opted to save money by storing the spent fuel on site rather than invest in safer storage options.

5.5 Nuclear Waste Disposal and Decommissioning Nuclear Plants

Radioactive waste is created at all points in the nuclear fuel cycle: from uranium mining, fuel enrichment and discharges from plants, to the highly radioactive waste resulting from reprocessing spent fuel and decommissioning contaminated sites.[130] The disposal of radioactive waste is one of the most difficult problems currently facing the nuclear power industry. High level waste and spent fuel disposal pose particularly acute problems given that it is the most toxic, long-lived and life-endangering wastes known to human kind.[131]

Earlier global planners did not consider the problem of nuclear waste until decades after nuclear power plants began operating. Perhaps oddly, the International Atomic Energy Agency did not hold its first meeting on decommissioning and permanent waste storage until 1973 – 20 years after the first reactor was built.[132] The waste problem is so technically difficult to handle and socially intractable with the public that in the past some countries got around the issue by dumping nuclear waste into the sea. France, for example, from 1967 to 1969 dumped more than 12 000 cubic meters of nuclear waste from the reprocessing plant at Marcoule into the ocean.[133] The ocean dumping of low level radioactive waste began in 1946 and took place in 50 different sites in the Atlantic and Pacific oceans,[134] but did not gather a great deal of disquiet until the 1970s, and was not halted until 1982 as a result of an international agreement.[xlviii]

From the 1970s onward, attempts were made to find a suitable way of dealing with radioactive waste – a problem highlighted in the sixth report of the UK Royal Commission on Environmental Pollution (RCEP) in 1976. One of the key recommendations to emerge from the so-called "Flowers Report" as it became known (after the chairman, Sir Brian Flowers) was that the UK should not embark on a programme of new nuclear power plants unless the question of waste disposal had been resolved. Waste slowly emerged as nuclear power's "Achilles' Heel".[136]

[xlviii] The LLW was usually packaged in metal drums lined with a concrete and bitumen matrix. As one reviewer observes, "So far, samples of sea water, sediments, and deep sea organisms collected on the various sites have not shown any excess in the levels of radionuclides above those due to nuclear weapons fallout, except on certain occasions where caesium and plutonium were detected at higher levels in samples taken close to packages at the dumping site".[135]

Across a number of countries the failed attempt to find a site for the geological storage of nuclear waste initiated a period of reflection on the part of the nuclear industry and its governmental backers. As a result, a newer, tentative and more open governance style was proposed whereby the formerly closed, secretive decision-making process was opened up. Previously excluded stakeholders were drawn into the process with mixed results. This required a culture change within organisations and a search for inclusive democratic processes that could enable debate to occur between previously antagonistic groups.[137,138] It was clear that without a solution for the long-term disposal of high level legacy wastes no new nuclear stations could be contemplated, a position reflected in many countries policy statements on nuclear energy.

Taking the UK as an example, after the rejection of a proposal to build a rock laboratory at Sellafield in order to test whether a geological disposal site would be geologically appropriate, the nuclear industry and government undertook a period of reflection that produced the Managing Radioactive Waste Safely (MRWS) process launched in 2001, which recognised that the closed decision-making process in the past had failed and sought a new open way forward based on stakeholder dialogue and deliberation.

As part of this "new transparency", a new committee, the Committee on Radioactive Waste Management CoRWM was ushered into existence in 2003 and was composed of people from scientific, technical and social scientific backgrounds. The committee was novel in its plural composition and its ambition to integrate scientific analysis with public and stakeholder engagement (PSE). CoRWM was to inspire public trust in decision making, which had suffered not only because of the failures in nuclear waste policy, but also due to previous incidents such as the BSE crisis.[139,140] The committee operated for three years, in which time it undertook the most ambitious public and stakeholder engagement process ever seen in the UK to date. In its final report in July 2006, one of the recommendations was to move forward with deep geological disposal of nuclear waste, but simultaneously called for an accompanying robust programme of research on interim storage and further R & D on deep geological storage.[141] These findings mirror those of comprehensive analyses published elsewhere which suggest that the science underpinning long-term geological isolation is sound and that the deep geological storage of high level waste is the most appropriate option.[142] Cost estimates for such a facility vary widely with a median figure of £12 billion. Whilst the majority of stakeholders supported the recommendations, a number of stakeholders (such as Greenpeace) and devolved administrations (Scotland) rejected them in favour of above ground interim storage. A recent report develops the criticisms of CoRWM recommendations.[143]

The CoRWM process illustrated what can be achieved when previously antagonistic stakeholders work collectively on a common problem, with sufficient time, resources and good will.[xlix] Whilst a move toward greater openness

[xlix] The experience of working on previous stakeholder dialogue projects which began during the mid-1990s in the nuclear arena has persuaded people within and outside of the nuclear industry of the possibility and potential of this form of collaborative working.

and engagement became evident in relation to finding a solution to long-term nuclear waste in a number of countries,[144] the picture in relation to pressing forward with new nuclear build is less reassuring. Contemporary policy discourse with its emphasis upon securitisation, in this case, energy security, reintroduced a policy making style redolent of early nuclear policy making. As Blowers observes, "the style of governance is less inclusive and participative", in many ways reverting back to some of the characteristics of what Dryzek describes as the "actively exclusive state" that had been prevalent in the UK until the early 1990s.[145] Whilst the government consultation during 2007 on new nuclear stations emphasised public participation, it was beset by a host of problems that led to it being successfully challenged in the courts by Greenpeace UK. The judge in a damming verdict stated that the consultation was "seriously flawed" and "manifestly inadequate and unfair" given that insufficient and "misleading" information had been made available by the government for consultees to make an "intelligent response".[146,1]

Similarly in France, the "Bataille Law" of 1991 on radioactive waste management marked a step towards a more democratic decision-making process, designed to put an end to the "cult of secrecy" that had hitherto prevailed in nuclear policy questions, but also to facilitate the exploration of different policy options.[148] European legislation on transparency and citizen participation also pushed the French nuclear establishment towards more openness. Whilst this new processes of consultation and discussion was beset by a number of problems, researchers have analysed the new arenas where experts and the public come together as revealing the limits of traditional representative democracy, suggesting the need to press forward with this emerging form of "technical democracy"[149] which, its been suggested, can increase public confidence and trust in the technical and organizational effectiveness of waste management and disposal.[150]

Although the cost of finding a solution to nuclear waste is eye watering, this is dwarfed by the price tag attached to the decommissioning of nuclear plants. Unlike a coal-fired power station, for example, a nuclear power plant cannot just be dismantled and the site used for other purposes. A complex process is initiated where radioactive parts, buildings and, on occasion, contaminated land must be carefully dismantled, treated and stored as nuclear waste. None of which is cheap. It is estimated that the final bill for decommissioning the UK's current fleet of nuclear plants will be in excess of £100 billion.[151] Much of the cost is attributable to the clean up of the oldest civilian and military plants.[152] Until recently, most cost estimates of nuclear plants excluded decommissioning costs, which can equal or exceed construction costs.[153] As a result the UK government is suggesting a levy will be imposed on electricity produced from

[1]The Blair administration began openly flagging new nuclear stations as a solution to climate change from 2006 onwards, despite CoRWM's clear statement that its conclusions should not be taken as either green or red light to new build; the alleged "solution" to the waste issue provided by CoRWM was used by the government in its arguments in favour of new build. This suggests that government can selectively and strategically deploy more open and transparent forms of decision making.[147]

nuclear plants to cover decommissioning and other back end costs, although there are concerns that this levy will not be sufficient.

For years the nuclear industry deferred decommissioning until as far as possible into the future. Even an official report published as late as 1995 suggested operators of nuclear plants in the UK could defer decommissioning for 100 years.[154] It was not until the decommissioning of the Berkeley Magnox reactors had to be planned after they stopped operating in 1988 and 1999 that the problem of how to pay for decommissioning came to light.[li] In coming to power in 1997 the New Labour administration began the process of searching for cost effective solutions to the decommissioning challenge, drawing on the experience of the US decommissioning effort in particular, which was based on contracting out decommissioning to private sector consortia. Rather than sell nuclear liabilities the government has "contractorised" them. As a result of the involvement of the private sector in more accurately capturing the true cost of decommissioning, liabilities increased by 16% alone in 2007.[156] The hope is that the skills and experience of the private sector will lead to the use of innovative solutions thereby driving down costs and reducing the final bill to the taxpayer. This has yet to be proven.

For most countries the preferred approach is to dispose of nuclear waste in facilities built in rock formations hundreds of meters below the earth. To date only Finland is in the process of building such an underground disposal facility, with Sweden only having recently come to an agreement with a local community as to where the facility will be built. On the other hand the country with the most nuclear waste, the USA, has just rejected a disposal site after investigating it for 20 years. For campaigners this seriously impacts upon the justification to expand the new build programme. For scientific and political reasons the Yucca Mountain site in Nevada, part of the Nevada nuclear test site, was chosen to store the nation's radioactive waste (a decision ratified by Congress) after a period of intensive research and debate in which $3 billion was spent. It was set to open in 2010 but opposition by environmental groups and Nevada politicians have kept things on hold. The Obama administration cancelled the plans after coming to power. The USA now has nowhere to place the 70 000 tonnes of waste currently being stored on-site at nuclear power plants and other facilities scattered throughout the country.[157]

5.6 *Proliferation Risks*

> *"One may suppose how radium could become very dangerous in criminal hands and here we might ask ourselves if it is to mankind's advantage to know the secrets of nature, if we are mature enough to profit from them or if that knowledge will harm us".*[158]

[li]The CEGB had claimed that funds for decommissioning were being set aside and that when decommissioning was to start the appropriate technology would be in place. However, it transpired that such funds were just a "bookkeeping exercise" with the money having been given back to HM Treasury, with no obvious mechanism for their return to fund decommissioning.[155]

Concerns over the potential malevolent use of nuclear materials have been on-going since the discovery of radiation over 100 years ago. In the present day, potential threats from malfunctions in nuclear reactors and/or from nuclear waste pale in comparison to the threats posed by the proliferation of nuclear materials and nuclear weapons. The series of threats never vanished with the end of the Cold War but continues to haunt us today, if not with more urgency than during the cold war itself. Nuclear weapons proliferation has been a concern since the birth of nuclear energy, given that the very purpose of the first nuclear reactors was to extract plutonium from the spent fuel for nuclear weapons. By contrast, today the objective is to minimize the proliferation risks of nuclear fuel cycle operation.[159]

The Treaty on the Non-Proliferation of Nuclear Weapons (NPT) came into effect in 1968. It stated that those states that already possessed nuclear weapons should not transfer atomic weapons to "non nuclear weapons states". The NPT also sought to invoke the discourse of human rights and development to justify nuclear energy, such that it was "the inalienable right of all the Parties to the Treaty to . . . use nuclear energy for peaceful purposes".[lii] Mohamed El Baradei, Director General of the IAEA, labels the enrichment and reprocessing capabilities of countries the "Achilles' heel" of the non-proliferation regime,[160] given that countries which possess such technologies have a virtual weapons programme.[liii] This is increasingly a problem as countries adapt or express an interest in developing Fast Breeder Reactors. The shortcomings of the NPT have led some observers to question the logic of a worldwide nuclear renaissance. Why is the expansion of nuclear energy a potential problem? The PUREX method of extracting plutonium from spent fuel is well known and easily accessible. At present there is 1000 tonnes of plutonium worldwide.

Countries which adopt PUREX/MOX may neither have the infrastructure or funds to control its spread. That said, there are real, though not insurmountable challenges for "rogue states" once they have acquired weapons grade material to actually develop weapons technology, which is much more of a technical challenge than commonly assumed[162] One solution is for the USA and other nuclear supplier group countries to lease fuel to countries with small nuclear programmes. A recent overview of the functioning of the NPT concluded that:

> *"The somewhat frayed non-proliferation regime will require serious re-examination and strengthening to face the challenge of the global growth scenario, recognizing that fuel cycle associated proliferation would greatly reduce the attraction of expanded nuclear power as an option for addressing global energy and environmental challenges".*[163]

[lii] The NPT is seen to have three pillars: (1) non proliferation, (2) disarmament, and (3) the right to peaceful nuclear technology.

[liii] The NPT is seen by some to encapsulate a "nuclear orientalism" in which "nuclear weapons are represented so that *theirs* are a problem whereas *ours* are not".[161]

6 Conclusions

We have seen through the birth, expansion, decline and renaissance of nuclear power, a number of claims that have sustained interest in the technology: from being too cheap to meter, to being means of securing energy independence and energy security, and as a solution to climate change. However, each of these claims has also had to deal with the challenges posed by the relatively high cost of building nuclear plants; persistent concerns over its safety to the perennial problem of finding a solution and site for the disposal of nuclear waste; as well as concerns over the proliferation risks inherent in the fuel cycle.

Part of the appeal of nuclear power today is the need to develop and foster low carbon energy sources in order to help mitigate climate change, coupled with this is the realisation that renewable energy sources (wind and solar) are not coming online quickly enough to enable a straightforward transition from a carbon dominated energy regime to a low carbon energy regime. As a result countries have either opted for a new nuclear build programme, sometimes after years of stagnation, or have opted to extend the lifetime of existing reactors. Whilst there has been relative stagnation in Europe and North America, a number of developing countries are embracing nuclear power not just for environmental reasons but as an integral part of their industrialisation process, to cope with the extraordinary growth in demand for electricity, a problem acutely facing China and India. The case for nuclear power also increases in parallel to arguments for the electrification of the transport and home/business heating sectors, again to mitigate climate change through development of a low carbon transport system. With technological developments such as the electric-powered car and bus becoming central to decarbonising the transport sector, carbon-free sources of electricity production will be privileged, with renewables (primarily wind) and nuclear power heading the list of sources. There are other non-economic factors at work in the desire for nuclear power, namely national status and prestige. Nuclear power becomes a means of political and social policy, strengthening the political legitimacy of the state in developing nations in particular.[164]

Given the overall resource limits and challenges we have identified, however, nuclear fission may only be able to play a short- to medium-term role in meeting these environmental and economic needs, as an enabling technology with a so-called "bridging role".[165] Even then, nuclear energy will face some significant problems as identified above, many of which have not yet been adequately addressed. Policymakers have designed a host of measures to try and address the more pressing concerns, such as creating loan guarantees, tax breaks and artificially raising the cost of carbon, all in an attempt to make nuclear more competitive in comparison with fossil fuels. We have shown that from its inception, nuclear energy has needed state support either directly through state funds or indirectly by intervening to structure markets in their favour.

A range of economic, technical and social challenges continue to plague the industry. Waste disposal is a prime example. Only in Finland is a deep geological facility under construction after years of discussion and debate with local communities, here public acceptance and trust is vital, with trust being more of a scarce commodity in some countries than others. The US waste disposal policy is in considerable disarray with profound disagreements among residents in Nevada, the proposed site of the storage facility. The problem of waste and spent fuel disposal is linked to the additional concern of proliferation. Given the weaknesses of the non-proliferation treaty, the expansion of nuclear power, particularly to countries with no history of nuclear energy, and the ability of states to use the PUREX process to extract plutonium, this is particularly worrisome. There are legitimate concerns that the expansion of nuclear power may also raise the risk of a serious accident in the core of a reactor, risks which are lessened given the improved design of the current generation of LWRs being built. However as we discussed above such risks are still too high and additional design changes to the LWR as well as alternative reactor designs must be developed to reduce the risk still further to within acceptable levels.

As the recent disaster at the Fukushima nuclear plant illustrates it is not just reactor design that needs to be improved but also regulatory oversight and senior management practises. We have seen at Fukushima how neglect of maintenance and safety rules can contribute to prolonging and intensifying the consequences of a natural disaster. Such consequences have been exacerbated by the failures of regulation in an age of privatization and the downsizing of government, as well as the inevitable, prosaic failure of organizations.[172] The building of new nuclear plants that are currently underway is unevenly geographically distributed but the challenges that states and corporations face are similar, the speed and scale at which the "nuclear renaissance" will occur depends on whether new reactors can demonstrate "better economics, improved safety, successful waste management, and low proliferation risk, and if public policies place a significant value on electricity production that does not produce CO_2".[166]

References

1. S. L. Montgomery, *The Powers That Be. Global Energy for the Twenty First Century and Beyond*, University of Chicago Press, Chicago and London, USA and UK, 2010.
2. W. J. Nuttall, *Nuclear Renaissance: Technologies and Policies for the Future of Nuclear Power*, IOP Publishing, Bristol, UK, 2005.
3. Department of State, *The Report on the International Control of Atomic Energy*, Publication 2498, 1946 (also known as the *Acheson–Lilienthal Report*).
4. R. Rhodes, *The Making of the Atomic Bomb*, Simon & Schuster, New York, 1998.
5. S. Cooke, *In Mortal Hands: A Cautionary History of the Nuclear Age*, Black Inc., Bloomsbury, USA, 2009.
6. W. C. Patterson, *The Technological Demands of Nuclear Power*, presented at the Nuclear Power and the Energy Future Symposium sponsored by the UK Atomic Energy Authority and Friends of the Earth, 1977.

7. A. Ham and R. Hall, A Way Forward for Nuclear Power, *Energy Rev.*, submitted.
8. S. M. Cohn, *Too Cheap to Meter. An Economic and Philosophical Analysis of the Nuclear Dream*, State University of New York Press, Albany, NY, USA, 1997.
9. Cooke 2009, *op. cit.*
10. World Nuclear Association, *Story of Nuclear Energy*, 2010e, http://www.world-nuclear.org/info/inf54.html 2010 (accessed 1st December, 2010).
11. J. Scurlock, in *Nuclear or Not?*, ed. D. Elliott, Palgrave, London, UK, 2007.
12. Cohn 1997, *op. cit.*
13. H. Wasserman, N. Solomon, R. Alvarez and E. Walters, *Killing our Own: the Disaster of America's Experience with Atomic Radiation*, Delacorte Press, New York, NY, 1982.
14. Cooke 2009, *op. cit.*
15. G. Gaivoronskaia and K. E. Solem, *Foresight*, 2001, **3**, 1.
16. D. Ford, *The Cult of the Atom*, Simon & Schuster, New York, NY, 1982.
17. World Nuclear Association, *Nuclear Power in the United Kingdom*, http://www.world-nuclear.org/info/inf84.html (accessed 23rd September 2010).
18. Cohn 1997, *op. cit.*, pp. 18–19.
19. R. Hawley, Nuclear power in the UK – Past, Present and Future, *Proceedings of the 31st World Nuclear Association Annual Symposium, Building the Nuclear Future: Challenges and Opportunities*, 2006.
20. Ham and Hall 2006, *op. cit.*
21. W. C. Patterson, *Nuclear Power*, Penguin Books, London, UK, 1976.
22. Hawley 2006, *op. cit.*
23. I. Jackson, *Nukenomics: The Commercialisation of Britain's Nuclear Industry*, Nuclear Engineering International Special Publications, 2008.
24. Scurlock 2007, *op. cit.*
25. Jackson 2008, *op. cit.*
26. T. Benn, *Nuclear Power and the Bomb*, http://www.tonybenn.com/nucl.html (accessed 1st December 2010).
27. Scurlock 2007, *op. cit.*
28. Jackson 2008 *op. cit.*, p. 6.
29. N. Fontaine, *Regards sur l'Actualité*, fév 2006, no. 318.
30. J. Fagnani and J.-P. Moatti, *J. Policy Anal.Manage.*, 1984, **3**(2).
31. M. Lehtonen, *J. Integrat. Environ. Sci.*, 2010, **7**, 3.
32. G. Hecht, *The Radiance of France: Nuclear Power and National Identity after World War II*, MIT Press, Cambridge, MA, USA, 1998.
33. J. Scurlock, Nuclear energy: an introductory primer, in *Nuclear or Not?*, ed. D. Elliott, Palgrave, London, UK, 2007.
34. A. Verbruggen, *Energy Policy*, 2008, 36.
35. *Public Utilities Fortnightly*, 1975, **95**(9), April 24.
36. V. Covello, *The Politics of Nuclear Power: A History of the Shoreham Nuclear Power Plant*, Kluwer Academic Publishers, 1990.
37. P. Högselius, Challenging Chernobyl's Legacy: Nuclear Power Policies in Europe, Russia and North America in the early 21st century, in *The*

Politics of Nuclear Energy in Asia, ed. Y.-C. Xu, Palgrave MacMillan, London, UK, in press.
38. C. Hadjilambrinos, *Energy Policy*, 2000, **28**.
39. C. Hadjilambrinos 2000, *op. cit.*
40. R. Williams, *The Nuclear Power Decisions: British Policies 1953–78*, London, UK, 1990.
41. T. Hall, (1986), *Nuclear Politics. The History of Nuclear Power in Britain*, Harmondsworth, UK, Penguin, 1966.
42. Shore cited in P. McAuslan, *Urban Law and Policy*, 1979, **2**(2), 1–23.
43. W. C. Patterson, *Going Critical: An Unofficial History of British Nuclear Power*, Paladin, London, UK, 1985.
44. Y. Zhou, C. Rengifo, P. Chen and J. Hinze *et al.*, *Energy Policy*, 2011, **39**, 771.
45. Zhou 2011, *op. cit.*
46. C. X. Liu, Z. Y. Zhang and S. Kidd, *Nucl. Eng. Design*, 2008, **238**.
47. W. C. Patterson, *Environment*, 1972, Dec.
48. Cooke 2009, *op. cit.*
49. Scurlock 2007, *op. cit.*
50. Scurlock 2007, *op. cit.*
51. G. Weightman, *New Society*, 1979, 8 Nov.
52. Högselius in press, *op. cit.*
53. A. Verbruggen, *Energy Policy*, 2008, **36**.
54. R. P. Ellis and M. B. Zimmerman, *Rev. Econom. Statistics*, 1983, **65**(2).
55. M. Freeman, *Execut. Intell. Rev.*, 2006.
56. F. E. Bonner, *Utilities Policies*, 1992, **2**(4).
57. P. Hoikka, Energia-asennetutkimus 1983 cited in M Lehtonen 2010, *op. cit.*
58. Kiljunen, 1986 cited in Lehtonen 2010, *op. cit.*
59. Wakeham cited in T. Stenzel, *What Does it Mean to Keep the Nuclear Option Open in the UK?* Parliamentary Office of Science and Technology Report E-13, 2003.
60. M. Tweena, *Nuclear Energy – Rise, Fall and Resurrection*, CICERO Working Paper, Oslo, Norway, 2006, 01.
61. Rough 2006, *op. cit.*, p. 17.
62. Hawley 2006, *op. cit.*
63. Jackson 2008, *op. cit.*
64. W. J. Nuttall, *Nuclear Renaissance: Technologies and Policies for the Future of Nuclear Power*, IOP Publishing, Bristol, UK, 2005.
65. B. N. Kursunoglu, S. Mintz and A. Perlmutter, *Preparing the Ground for Renewal of Nuclear Power*, Plenum Press, 1999.
66. A. Blowers, in *Nuclear or Not?*, ed. D. Elliott, Palgrave, London, UK, 2007.
67. J. Lovelock, Nuclear Power is the Only Green Solution, *The Independent*, 2004.
68. J. Lovelock cited in A Ma'anit. A, *New Internationalist*, Sep. 2005.
69. G. Boyle, A tale of two countries: non-nuclear sustainable energy futures for the Germany and the UK, in *Nuclear or Not?*, ed. D. Elliott, Palgrave, London, UK, 2007.
70. IPCC, *Climate Change 2001: The Scientific Basis*, Cambridge Univ. Press, Cambridge, UK, 2001.

71. Carbon Emissions per Person, by Country, *Guardian Online*, http://www.guardian.co.uk/environment/datablog/2009/sep/02/carbon-emissions-per-person-capita (accessed on 1st December 2010).
72. *The Future of Nuclear Power: An Interdisciplinary MIT Study*, MIT, 2003.
73. S. Kyllönen, Ydinvoiman ilmastonmuutos. In Kojo, ed. M. Ydinvoima, valta ja vastarinta, Keuruu, cited in Lehtonen 2010, *op. cit.*
74. L. E. Echavarri, *Electricity J.*, 2008, **20**(9).
75. B. K. Sovacool, *J. Integrat. Environ. Sci.*, 2010, **7**(2).
76. M. Nicholson *et al.*, *Energy*, 2011, **36**(1).
77. N. Stern, *The Economics of Climate Change*, Cambridge University Press, Cambridge, UK, 2007.
78. Lehoten 2010, *op. cit.*
79. Zhou 2010, *op. cit.*
80. Survey results published by Deloitte & Touche LLP on 2 December 2005.
81. Lehtonen 2010, *op. cit.*
82. World Nuclear association 2010, *op. cit.*
83. P. Kiljunen, http://www.sci.fi/~yhdys/eas_08/sisallys.htm (accessed 1st December 2010).
84. K. Bickerstaff *et al.*, *Publ. Understanding Sci.*, 2008, **17**(2).
85. M. Schneider, *Bull. Atom. Scientists*, 2008, 3 Jun.
86. P. Jackson 2008, *op. cit.*, p. 95.
87. S. Kidd, in *Nuclear or Not?*, ed. D. Elliott, Palgrave, London, UK, 2007.
88. Zhang 2009, *op. cit.*
89. Zhou 2010, *op. cit.*
90. S. Reiss, Let a Thousand Reactors Bloom, *Wired Magazine*, 2004.
91. X. Yi-chong, *The Politics of Nuclear Energy in China*, Palgrave, London, UK, 2010.
92. W. Dazhong and L. Yingyun, *Nucl. Eng. Design*, 2002, **218**.
93. R. Heinberg, *The Ecologist*, 2008, 9th Jun.
94. G. T. Miller and S. Spoolman, *Living in the Environment: Principles, Connections and Solutions*, Brooks/Cole, 2009.
95. *Wall Street Journal*, 2010, 16th Nov.
96. Zhou 2010, *op. cit.*
97. World Nuclear Association, *Supply of Uranium*, (accessed 1st December 2010). www.world-nuclear.org/info/inf75.html
98. Montgomery 2010, *op. cit.*
99. WNA 2010, *Supply of Uranium, op. cit.*
100. WNA 2010, *Supply of Uranium, op. cit.*
101. Uranium Information Center, *Nuclear Electricity*, 6th edn, 2000, ch. 3, www.uic.com.au/ne3.htm (accessed 1st December 2010).
102. B. L. Cohen, *Am. J. Phys.*, 1983, **51**(1).
103. MIT, *The Future of the Nuclear Fuel Cycle: An Interdisciplinary Study*, 2010.
104. Montgomery 2010, *op. cit.*
105. MIT 2003, *op. cit.*
106. MIT 2003, *op. cit.*
107. Montgomery 2010, *op. cit.*

108. Polson and Katz, Constellation Drops Nuclear Plant, Denting EDF's U.S. Plans, *Bloomberg Online*, 2010, http://www.bloomberg.com/news/2010-10-09/constellation-drops-maryland-nuclear-plant-loan-denting-edf-s-u-s-plans.html (accessed 20th December 2010).
109. S. Wald, *Scientific American Earth 3.0;* Dec Special Edn, 2008, 18, 5.
110. IME 2010, *op. cit.*, p. 3.
111. IME 2010, *op. cit.*
112. Chris Huhne unveils plans for reform of UK energy market, *The Guardian*, 2010, 16th Dec.
113. Ramana 2009, *op. cit.*
114. Sovacool 2010, *op. cit.*, p. 107.
115. A. Frogatt, in *Nuclear or Not?*, ed. D. Elliott, Palgrave, London, UK, 2007.
116. Nuclear Power International 2010, *op. cit.*
117. Jackson 2008, *op. cit.*, p. 105.
118. Institute of Mechanical Engineers, *Nuclear new build. A vote of no confidence?* 2010.
119. World Nuclear Association, *The Economics of Nuclear Power*, http://www.world-nuclear.org/info/inf02.html (accessed 1st November 2010).
120. New nuke plants face skilled labor shortage, *CBS News*, 2010, Apr 30.
121. Schnider 2008, *op. cit.*
122. Schneider 2008, *op. cit.*
123. Y. J. Guo, *Perspectives on China's Nuclear Working Force*, 2004, http://www.caea.gov.cn/n16/n1223/46586.html (accessed on 1st January 2010).
124. G. J. Li and Z. Z. Ding, *Chin. J. North China Electric Power Univ.*, 2006, 1.
125. Sovacool 2010, *op. cit.*, p. 108.
126. Ramana 2009, *op. cit.*
127. MIT 2003, *op. cit.*, ch. 6.
128. MIT 2003, *op. cit.* p. 49.
129. MIT 2003, *op. cit.*, p. 47.
130. Frogatt 2007, *op. cit.*
131. J. Byrne and S. Hoffman, *Governing the Atom: The Politics of Risk*, Transactions Publishers, 1996.
132. Raffensperger 1996, *op. cit.*
133. Greenpeace, *Talking Nonsense – The 2007 Nuclear Consultation*, 2007, Sep, www.greenpeace.org.uk/files/pdfs/nuclear/2007-consultation-nuclear-dossier.Pdf (accessed 1st November 2010).
134. D. P. Calmett, *Int. Atomic Energy Authority Bull.*, 1989, 4.
135. Calmet 1989, *op. cit.*, p. 47.
136. Lowry 2007, *op. cit.*
137. A. Bergmans, M. Elam, D. Kos, M. Polic, P. Simmons, G. Sundqvist and J. Walls, *Wanting the Unwanted: Effects of Public and Stakeholder Involvement in the Long-term Management of Radioactive Waste and the Siting of Repository Facilities*, Final Report CARL Project, 2008.
138. A. Blowers, *J. Integrat. Environ. Sci.*, 2010, **7**(3).
139. CoRWM 2006, *op. cit.*
140. Chilvers and Burgess 2008, *op. cit.*

141. CoRWM, *Managing Our Radioactive Waste Safely*, 2006.
142. MIT 2003, *op. cit.*
143. Wallace 2010, *op. cit.*
144. Blowers 2010, *op. cit.*
145. Blowers 2010, *op. cit.*, p. 166.
146. The Queen on the application of Greenpeace Limited -v- Secretary of State for Trade and Industry.
147. Blowers 2010, *op. cit.*
148. Y. Barthe, *Le Pouvoir d'Indécision. La Mise en Politique des Déchets Nucléaires*, Economica, Paris, France, 2006.
149. M. Callon, P. Lascoumes and Y. Barthe, *Acting in an Uncertain World: An Essay on Technical Democracy*, MIT Press, 2009.
150. MIT 2003, *op. cit.*
151. Jackson 2008, *op. cit.*
152. Kidd 2007, *op. cit.*
153. Raffensperger 1996, *op. cit.*
154. Jackson 2008, *op. cit.*
155. Patterson 1990, *op. cit.*
156. Jackson 2008, *op. cit.*
157. von Hippel 2008, *op. cit.*
158. P. Curie in 1905, cited Cooke 2009, *op. cit.*, p. 45.
159. MIT 2003, *op. cit.*
160. N. Dombey, *New Left Rev.*, 2008, Jul–Aug, 52.
161. Gusterson, *Cultural Anthropol.*, 1999, **14**, 1.
162. J. Mueller, *Atomic Obsession: Nuclear Alarmism from Hiroshima to Al-Qaeda*, Oxford University Press, Oxford, UK, 2010.
163. MIT 2003, *op. cit.*, p. 67.
164. Montgomery 2010, *op. cit.*
165. Jackson 2008, *op. cit.*
166. MIT 2009, *op. cit.*
167. Gallup Inc., http://www.gallup.com/poll/146660/disaster-japan-raises-nuclear-concerns.aspx.
168. U.S. Support for Nuclear Power Climbs to New High of 62%, http://www.gallup.com/poll/126827/support-nuclear-power-climbs-new-high.aspx.
169. J. Chaffin, Sharp rise in European anxiety on nuclear safety, 2011, http://www.ft.com/cms/s/0/a378cf98-759c-11e0-80d5-00144feabdc0.html#axzz1Lrl524A3.
170. Nucléaire: une trentaine de réacteurs dans le monde risquent d'être fermés', *Les Échos*, published 12th April 2011.
171. M. Schneider, A. froggatt and S. Thomas, *Nuclear Power in a Post-Fukushima World 25 Years after the Chernobyl Accident*, The World Nuclear Industry Status Report 2010–2011.
172. C. Perrow, *The Next Catastrophe*, Princeton University Press, Princeton, NJ, 2007, pp 133/134.
173. J. McCurry, Japan nuclear firm admits missing safety checks at disaster-hit plant, *The Guardian*, 22 March 2011.

Nuclear Fuel Cycles: Interfaces with the Environment

CLINT A. SHARRAD, LAURENCE M. HARWOOD AND
FRANCIS R. LIVENS*

ABSTRACT

The waste materials generated in the nuclear fuel cycle are very varied, ranging from the tailings arising from mining and processing uranium ore, depleted uranium in a range of chemical forms, to a range of process wastes of differing activities and properties. Indeed, the wastes generated are intimately linked to the options selected in operating the nuclear fuel cycle, most obviously to the management of spent fuel. An open fuel cycle implies the disposal of highly radioactive spent fuel, whereas a closed fuel cycle generates a complex array of waste streams. On the other hand, a closed fuel cycle offers options for waste management, for example reduction in highly active waste volume, decreased radiotoxicity, and removal of fissile material. Many technological options have been proposed or explored, and each brings its own particular mix of wastes and environmental challenges.

1 Nuclear Fission as an Energy Resource

The vast majority of nuclear reactors which operate, or have operated, have produced energy from uranium fuel. The neutron-induced fission of a uranium nucleus typically yields around 200 MeV of energy (compare *ca.* 4 eV per atom in the oxidation of carbon to CO_2), so the energy density of nuclear fuel is very high, and the volumes of fuel required and waste produced are relatively small.

* Corresponding author

Issues in Environmental Science and Technology, 32
Nuclear Power and the Environment
Edited by R.E. Hester and R.M. Harrison

Published by the Royal Society of Chemistry, www.rsc.org

Nevertheless, because of their radioactivity, wastes from nuclear energy production are potentially very hazardous over long timescales and their management is often both politically contentious and technically demanding. While there is considerable international cooperation in the area of nuclear energy, through for example the activities of the IAEA or the OECD's Nuclear Energy Agency, and international commercial activities such as fuel reprocessing and fuel manufacture do occur, the management of radioactive wastes is presently seen as being a national responsibility. The following discussion is largely written from a UK perspective, although it draws on examples from overseas where appropriate.

2 Nuclear Fuel

The early generations of nuclear fission technology depended on the thermal neutron (neutrons with an energy ~0.025 eV) fission of the natural isotope ^{235}U, present in nature at 0.72 atom%. The dominant uranium isotope, ^{238}U, is a so-called "fertile" isotope since it can be converted into artificial isotopes, especially the fissile ^{239}Pu, by neutron irradiation. Thermally fissile plutonium isotopes can also be exploited in energy production. So, as plutonium and other fissile isotopes are produced through irradiation of ^{238}U, they can also be exploited in energy production, either through consumption *in situ*, or through recycling into new fuel materials. In a uranium-fuelled thermal reactor, about 40% of total energy is derived from the fission of plutonium isotopes produced *in situ*.

2.1 *Uranium Mining*

There is a comprehensive discussion of uranium mining on the World Nuclear Association website (http://www.world-nuclear.org/), on which the following discussion is largely based. The first step in the nuclear fuel cycle is the mining of uranium ore. While high grade ores are still available, a significant proportion of uranium mining is now carried out by extraction of large volumes of easily accessed low grade ore (grade typically a few hundred ppm U_3O_8) from open cast mines. Canada, Australia, Kazakhstan, Niger, Russia and Namibia presently produce most of the World's uranium. World uranium production has increased by almost 50% over the last decade, to over 50000 tonnes in 2009.

As well as the naturally occurring uranium isotopes, ^{234}U, ^{235}U and ^{238}U, uranium ores contain a wide range of other radioisotopes, formed *in situ* as intermediates in the decay of uranium to stable lead isotopes. These are dominated by the decay products of ^{238}U (see Figure 1). Uranium is extracted from crushed ore by leaching, usually with either sulfuric acid or sodium carbonate solution, then concentrated from the leachate by solvent extraction or ion exchange. Most of the decay product radionuclides (Ra and below) are left in the wastes. Especially where ore grades are low, large volumes of these wastes ("tailings") arise and uranium mine wastes are often relatively radioactive. In total, around 940 Mt of tailings have been created.[1] These wastes

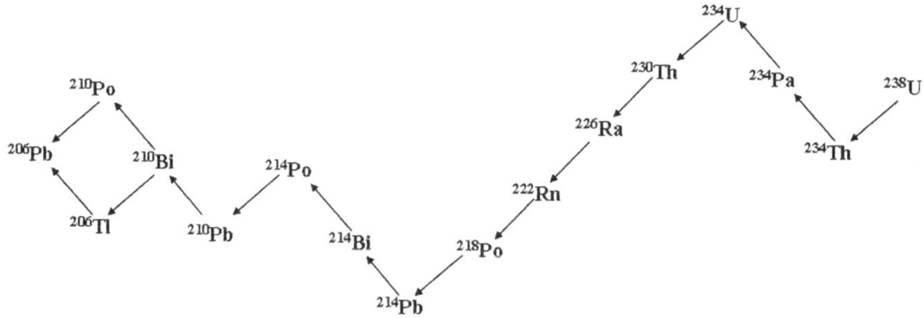

Figure 1 Decay products derived from ^{238}U. Downward arrows denote α-decays, while upward arrows denote β-decays.

require careful management to prevent the spread of contamination and associated health risks. A particularly challenging example is the Erzgebirge of eastern Germany (http://www.wise-uranium.org/uwis.html), where 216 000 tonnes of uranium were extracted between 1945 and 1990. These mining activities affected an area of about 100 km^2, primarily around five mine sites and two ore processing sites. The wastes included 311 million m^3 of waste rock, and a further 178 million m^3 of tailings. The tailings covered a total area of almost 600 hectares, to a maximum thickness of 70 m. A 15 year remediation programme, costing around €6 billion and largely now completed, has been required to stabilise and restore the area.

2.2 Uranium Fuel Production and Use

After extraction from ore, a uranium concentrate ("yellowcake") is manufactured and shipped to facilities where it is enriched (the proportion of ^{235}U is increased from the natural 0.72 atom%) if required and fabricated into fuel. Enrichment is needed for modern reactor fuels which are made from UO_2, and generally requires conversion of the uranium into UF_6, a relatively volatile compound which is attractive for enrichment because fluorine is monoisotopic, followed by multiple stages of membrane diffusion or centrifugation. The enrichment process, as used for fuel production, creates two uranium streams, one enriched, typically to 3–5 atom% ^{235}U (referred to as low enriched uranium), and one depleted to around 0.2 atom% ^{235}U. Typically, therefore, production of 1 kg of low enriched uranium creates 5 or 6 kg of depleted uranium, for which there is little current use. In total, about 1.2 M tonnes of depleted uranium exist worldwide. A significant proportion of the global depleted uranium inventory is still in the form of UF_6, a reactive, corrosive material which is not suitable for long term storage or disposal.

For use in current reactors, enriched uranium is "deconverted" from UF_6 into UO_3, reduced to UO_2, a durable ceramic, and formed into pellets. The pellets are loaded into metal tubes, generally of stainless steel or zircaloy (a range of zirconium-based alloys, often containing tin or niobium), depending on reactor type, and are then suitable for loading into a reactor. Early reactors, such as the first Hanford production reactors in the USA, and the UK Magnox reactors, were designed to use fuel of natural isotopic composition, which obviously avoids the difficulty and cost of enrichment. However, such reactors cannot tolerate the dilution of fissile isotopes which occurs in UO_2, and have to use uranium metal, whose properties limit reactor operating temperatures and efficiency.

2.3 Modern Civil Reactor Fuels

Two types of fuel are used in modern, commercial power reactors. Both are oxide materials, which are well suited to the demanding heat and radiation environment in the reactor. The more common is uranium, as UO_2, with the ^{235}U isotopic content typically enriched from the natural abundance to between 3 and 5 atom%. Occasionally, higher levels of enrichment, up to *ca.* 9 atom%, are used. Some reactors, notably in Japan, France and Switzerland, are partially loaded with mixed U/Pu oxide fuel (MOX). This typically comprises uranium with low ^{235}U content, either the depleted uranium byproduct of enrichment, or from recycling of used ("spent") uranium fuel, blended with plutonium to enhance its reactivity. MOX typically comprises a blend of 7% Pu/93% U. A modern pressurised water reactor (PWR) can accommodate core loadings of up to *ca.* 1/3 MOX without significant design implications.

2.3.1 Plutonium

The potential utility of plutonium as a component of MOX fuel depends on its isotopic composition. Thermal irradiation can produce isotopes from ^{239}Pu to ^{242}Pu inclusively (see Table 1). As irradiation time increases, so does the proportion of heavier isotopes, with mass number >239. Only ^{239}Pu and ^{241}Pu are fissile in a thermal reactor, so plutonium separated from high burnup fuels will have a lower fissile content than that separated from low burnup fuels. In addition, ^{241}Pu has a relatively short half life so, after storage periods of years to decades, a significant proportion will have decayed to ^{241}Am. This has

Table 1 Properties of plutonium isotopes. Data for different fuel types from NDA.[2] Magnox fuel has a natural isotopic composition with a burnup of 3000 MWd tonne^{-1}; AGR fuel and PWR fuel are low enriched fuels with burnups of 18000 MWd tonne^{-1} and 53000 MWd tonne^{-1}, respectively.

	Half Life (yrs)	*Decay*	*Atom% in Magnox Fuel*	*Atom% in AGR Fuel*	*Atom% in PWR Fuel*
^{238}Pu	87.7	α	0.1	0.6	2.7
^{239}Pu	24110	α	80.0	53.7	50.4
^{240}Pu	6563	α	16.9	30.8	24.1
^{241}Pu	14.35	β	2.7	9.9	15.2
^{242}Pu	3.73×10^5	α	0.3	5.0	7.1

two effects. First, the fissile content of the plutonium has decreased and, second, the decay of ^{241}Am is accompanied by a 59.5 keV gamma emission, which makes handling much more difficult because shielded facilities are needed to handle plutonium with a significant americium content. Thus, the Melox plant at Cadarache, which produces MOX fuel, is limited to using plutonium which is less than five years old.

The isotope ^{238}Pu is also formed in irradiated uranium. This is formed by neutron capture in ^{237}Np, itself produced through either an n,2n reaction in ^{238}U, or by successive neutron captures in ^{235}U.

2.4 Irradiation of Nuclear Fuel

As any fuel is irradiated, the proportion of useful fissile isotopes decreases and the content of fission products increases. Some of the fission products are efficient neutron absorbers, and "poison" the fuel. Consequently, after approximately three years irradiation, the reactivity of the fuel is too low and it is necessary to remove the "spent" fuel from the reactor and replace it with fresh. The composition of a typical uranium fuel is summarised in Table 2.

2.5 Alternative Fuels

2.5.1 Uranium/Plutonium Fast Reactors

The high energy ("fast") neutrons produced in fission have an energy >1 MeV and can fission ^{238}U effectively. In addition, the high energy fission events triggered by fast neutrons produce more neutrons (*e.g.* in ^{239}Pu, 4.9 for fast *versus* 2.6 for thermal) so that there is a substantial surplus of neutrons which can be used either to transmute a fertile material (a breeder reactor) or to destroy problematic waste isotopes. A thermal reactor is specifically designed to "moderate" neutron energies by allowing the neutrons to collide with

Table 2 Example compositions of fresh and spent nuclear fuels, excluding oxygen. It is assumed that the fresh uranium fuel is fabricated from unrecycled uranium. After irradiation, the isotopic composition of uranium is changed, with the ^{235}U content decreased, and production of ^{236}U from neutron capture. The composition of spent MOX is quite variable depending on fuel composition and irradiation history, so these data should only be viewed as approximations. MOX data from WNA.[3]

	Fresh Uranium Fuel (%)	*Spent Uranium Fuel (%)*	*Fresh MOX (%)*	*Spent MOX (%)*
Fission Products	0.0	3.4	0.0	4.7
Uranium	100	95.6	90–97 (typically 93)	90
Pu Isotopes	0.0	0.9	10–3 (typically 7)	5
Minor actinides (Np, Am, Cm)	0.0	0.1	0.0	0.3

light atoms. It is, however, also possible to design reactors where reaction is sustained by fast, not thermal, neutrons. While such reactors could be supported by variants on the present U/Pu fuel cycle, there are major engineering difficulties associated with them due, for example, to the very high energy density in the cores, and the need to use corrosive materials such as liquid metals as coolants. Technologically, fast reactors are much less well developed than thermal reactors.

2.5.2 Highly Enriched Uranium

Specialised reactors, primarily used for research or marine propulsion, use highly enriched uranium (HEU), which contains $>20\%$ ^{235}U, as fuel. This presents particular proliferation risks since the fuel contains weapons-usable uranium and, over the last 20 years, considerable efforts have been made to remove HEU from research reactors and to reconfigure them to use lower enriched fuel. HEU fuels typically contain lower plutonium contents than natural or LEU fuels, since they have a lower content of the ^{238}U precursor. As well as the high fissile content, HEU fuels often have unusual compositions and structures which make them difficult to reprocess in a conventional plant.

2.5.3 Thorium

The commonest isotope of thorium, ^{232}Th, is fertile, being converted by neutron irradiation to the fissile ^{233}U. However, there is no thermally fissile isotope of thorium available in nature in usable amounts, so it is not possible to construct an entirely thorium-fuelled reactor. Any thorium-fuelled reactor therefore has to use a fissile material such as ^{233}U, ^{235}U or ^{239}Pu to drive the reaction.

Thorium fuel has several advantages because the fissile ^{233}U is significantly contaminated with a ^{232}U byproduct formed by n,$2n$ reactions in ^{233}U. The ^{232}U decay products emit high energy gamma rays, which limits the utility of a $^{233}U/^{232}U$ mixture in nuclear weapons. In addition, long-lived transuranic wastes are much less significant in thorium fuels, although ^{231}Pa (half life 3.27×10^4 years) is produced. Disadvantages include the presence of ^{232}U and its decay products, which make handling of the uranium stream difficult, since that has to be conducted in heavily shielded facilities. Also, large scale recycling of thorium based fuels may require novel technologies.

3 Nuclear Fuel Recycling

As outlined earlier, a large proportion of spent uranium fuel is potentially reusable. The vast majority is still uranium which, although reduced in fissile content, contains residual enrichment and can be recycled. Plutonium can also be recovered for use in MOX or other fuels. In addition, recycling of used fuel may reduce waste volumes for disposal and/or allow removal of high hazard radionuclides for separate treatment. Separation of used fuel on an industrial scale is, however, a complex and challenging technology.

3.1 Separation of Uranium and Plutonium

The separation of plutonium from irradiated nuclear fuel was originally developed in the nuclear weapons programmes of the 1940s and 1950s. Since nuclear weapons require plutonium of high fissile (^{239}Pu) content, irradiation times in production reactors were short and burnups were low, no more than a few hundred MWd $tonne^{-1}$. In these feed materials, the fission product loadings were therefore also low, and the content of higher actinides was small, so that they presented a much less severe challenge to separations technology than modern, high burnup fuels from civil reactors. In addition, little attention was paid to waste management in early weapons' production programmes.

All large scale separations depend on the diverse redox chemistry which characterises the mid-actinides (see Table 3). In the media of interest for separations, the substantial chemical differences (for example in solubility in aqueous or non-aqueous solvents, or in affinity for a complexant) between the linear dioxo "actinyl" ions, $MO_2^{+/2+}$ formed by oxidation states V and VI, and the "simple" $M^{3+/4+}$ ions formed by lower oxidation states generally provide the basis of useful separations. Solvent extraction processes, which are the mainstay of current technology, often exploit the differing affinity of different actinide species for selective complexants, usually O-donor ligands. The following examples are not an exhaustive description but serve to illustrate the diversity of processes which have been explored or used, and of the waste streams which can be produced.

3.2 Other Reasons to Reprocess

In addition to its original purpose, of recovering uranium and plutonium, reprocessing also offers the possibility of:

(i) Controlling proliferation risks. The production and potential isolation of fissile materials which can be diverted from the civil fuel cycle for military purposes is an intrinsic risk in nuclear power. Although the fissile content of plutonium derived from high burnup fuel is not optimal, power reactor plutonium can nevertheless be used in a weapon,[4] so the creation of plutonium is, by definition, a proliferation risk. While irradiated fuel is lethally radioactive and specialised facilities would be

Table 3 Principal oxidation states of the mid actinides. Bold indicates oxidation states which are significant in nuclear fuel recycling.

Uranium	*Neptunium*	*Plutonium*	*Americium*
III	III	**III**	**III**
IV	**IV**	**IV**	IV
V	**V**	V	V
VI	**VI**	VI	VI
	VII	VII	

needed to recover fissile material, the fission products will decay over a few hundred years to the point where plutonium could be easily recovered. This raises complex ethical issues and the recovery of plutonium for separate treatment, either as a waste or as a fuel, within a few years of production can be attractive for these reasons.

(ii) Reducing waste volumes. Well over 90% by mass of spent fuel is uranium. If a "once through" or "open" fuel cycle is adopted, the irradiated fuel will be packaged and disposed as waste. As a result, the volume of waste for disposal will be very substantial, and the associated costs will be high. For example, in some disposal concepts being considered for the UK, fuel elements containing 2–4 tonnes heavy metal (masses of nuclear materials are often expressed in tonnes heavy metal (tHM), *i.e.* equivalent mass of uranium or plutonium in the material) depending on fuel type and heat production could be packaged in a cast iron insert, then in a copper container between 2 and 5 m long and 0.9 m diameter, with 5 cm thick walls.[5] By contrast, reprocessing spent fuel and conversion of the high level waste to glass will produce less than 100 kg (0.04 to 0.05 m^3) of glass per tonne of uranium reprocessed, reducing the volume of highly radioactive material for disposal. Since high level waste and spent fuel are heat generating wastes, they need to be widely spaced in a disposal facility to limit the heat load, so the volume of waste disposed has a large effect on the facility footprint and consequent cost. If there is only a limited volume of host rock, reducing waste volume may be very helpful. Finally, a disposal facility, for example the currently suspended Yucca Mountain facility in the USA, may be legally limited to a specific volume of waste, in which case volume reduction by removal of uranium may well be attractive.

(iii) Controlling high level waste radiotoxicity. The majority of the fission products in spent fuel have relatively short half lives, so that, at timescales longer than a few hundred years, the activity is dominated by relatively radiotoxic actinide elements (see Figure 2). If, in addition to conventional reprocessing to remove uranium and plutonium, a further separation of minor actinides (*e.g.* neptunium, americium and curium) from high level waste is carried out, the radiotoxicity of the waste can be reduced by several orders of magnitude beyond a thousand years or so. Of course, the concentrated minor actinide stream has to be managed separately, which prompts much of the current interest in transmutation processes.

3.3 Historical Reprocessing Technologies

3.3.1 Precipitation Processes

The earliest separations were developed in support of the WW2 Manhattan Project and exploited the different solubilities of Pu(IV) and U(VI). The primary centre for these activities was the Hanford site in Washington State, USA.[7] Most fission products are soluble in strongly acid conditions, whereas the fluorides and phosphates of the tri- and tetravalent actinides are not.

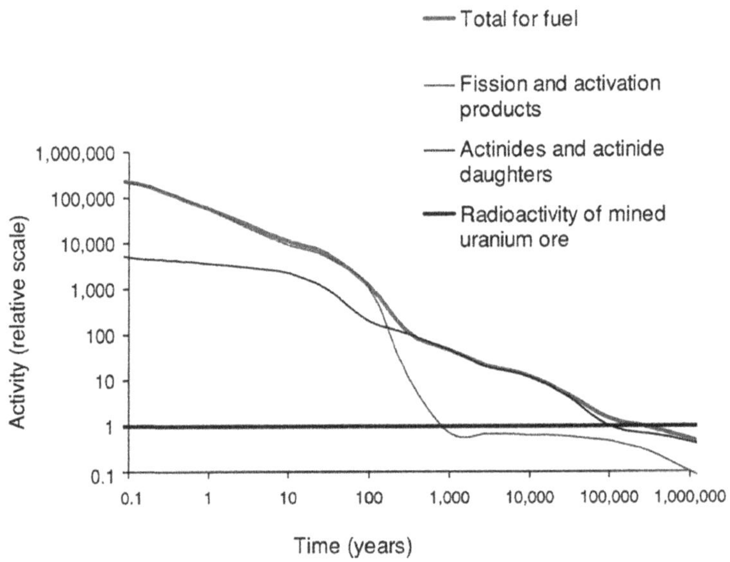

Figure 2 Radioactivity of irradiated fuel components relative to uranium ore as a function of time.[6]

Thus, these separations were based on dissolution of fuel in nitric acid, with control of the uranium oxidation state to VI and of the plutonium oxidation state to III or IV. On addition of Bi, followed by phosphoric acid, $BiPO_4$ was precipitated, carrying plutonium and separating it from uranium and many fission products. The crude precipitate was then dissolved in strongly oxidising conditions, for example with BiO_3^{2-} or MnO_4^- in the presence of $Cr_2O_7^{2-}$ holding oxidant, which switched the plutonium to oxidation state VI. Further purification was achieved by lanthanum fluoride precipitation, which eliminated lanthanide fission products and isolated plutonium.

These processes were carried out on an industrial scale from the end of 1944 until the early 1950s and were used to manufacture tonne quantities of plutonium from hundreds of tonnes of uranium (the irradiated fuel contained about 250 mg Pu $tonne^{-1}$ U). In the early 1950s, precipitation separation was replaced by solvent extraction, leading to the development of the Purex process and a substantial increase in throughput. By the early 1960s, Hanford was separating between 1.5 and 2 tonnes of plutonium from 7000 tonnes of irradiated uranium each year, 80% of this from the Purex process.

3.4 Purex

Purex (Plutonium URanium EXtraction) separation has been the predominant process used for industrial scale nuclear fuel reprocessing in France and the United Kingdom, two countries which have devoted much effort to closing the civil nuclear fuel cycle. In principle, Purex chemistry is based on solvent extraction and is straightforward. The irradiated fuel is dissolved in HNO_3 and the oxidation states of uranium and plutonium are controlled to VI and IV, respectively. The fuel solution is contacted with tri-*n*-butyl phosphate in an inert solvent, such as kerosene, and the uranium and plutonium are extracted into the solvent phase, leaving almost all fission products in the aqueous phase, which forms the high level liquid waste stream. The plutonium is then separated by reduction to oxidation state III and stripped into dilute HNO_3. Finally a uranium stream is separated by back extraction into dilute HNO_3.

3.5 Wastes from Fuel Reprocessing

In practice, Purex, like any chemical separation process, generates multiple waste streams ranging in activity from high level liquid waste to trace active washings. Solid wastes are also produced, comprising fuel cladding material (*e.g.* stainless steel, graphite and zirconium), as well as contaminated process plant components and miscellaneous industrial waste. Prior to reprocessing, the irradiated fuel is stored for several years (in the UK, usually in water-filled ponds) so the pond waters also form low level effluents.

Preparation for separation involves shearing and dissolution of fuel, which will lead to the release of volatile radionuclides, principally ^{3}H, ^{14}C, ^{106}Ru and ^{129}I. These may be trapped for subsequent immobilisation and disposal, or they may be released. Thus, nuclear fuel reprocessing produces a diverse range of solid, liquid and gaseous waste streams, all of which will impact on the environment; for example, the current UK Radioactive Waste Inventory identifies over 180 different waste streams currently being produced at Sellafield.[8] All of these waste streams have potential environmental impacts, and all have to be managed to ensure that their impacts are tolerable.

3.6 Other Solvent Extraction Processes

Many different solvent extraction processes have been devised to produce particular product and waste streams. For example, it may be attractive to avoid separation of a pure plutonium stream in order to limit proliferation risks, or it may be useful to separate long-lived isotopes from high level liquid waste for separate treatment, so that the radiotoxicity of the remaining high level waste decreases more rapidly through decay. In some cases, separations are designed to fit with specific national regulatory requirements and many of these processes could be used in combinations to give particular, desired outcomes. Some examples are given in Table 4.

4 Waste Management Options

The critical decision in the "back end" of the nuclear fuel cycle is whether or not to reprocess (in other words, whether the cycle is closed or open). If an open fuel cycle is chosen (*e.g.* as in Sweden), then waste management is essentially confined to the management of spent fuel and reactor decommissioning wastes,

Table 4 Examples of actual and candidate processes for the recycling of spent nuclear fuel.[14]

Process	*Purpose*	*Medium*	*Extractant(s)*	*Products*	*Example Reference*
TRUEX	Separation of transuranics from waste streams	Nitric or hydrochloric acids	– octyl (phenyl)-*N*,*N*-diisobutylcarbamoylmethylphosphine oxide (CMPO) – TBP – OK	– TRU – Waste stream disposable as non-transuranic waste	ref. 9
DIAMEX		Nitric acid	– Diamide (*e.g.* dimethyldibutyltetradecylmalonamide (DMDBTDMA)) for lanthanide + actinide separation, followed by separation of lanthanides from actinides with, for example alkylated tripyridyltriazine	– TRU – Ln	ref. 10
UNEX		Nitric acid	– Chlorinated cobalt dicarbollide – Polyethylene glycol – Diphenyl-*N*,*N*-di-*n*-butylcarbamoylmethyl phosphine oxide – phenyltrifluoromethyl sulfone diluent	– ^{137}Cs – ^{90}Sr Group separation of Ln & An	ref. 11
GANEX	All actinides from fission products	Nitric acid	– *N*,*N*,*N*′,*N*′-tetraoctyldiglycolamide (TODGA) – TBP – OK	Group separation of An from FP	ref. 12
SANEX	Separation of Am, Cm from FP in a purified HLW stream	Nitric acid	– 6,6′-bis(5,5,8,8-tetramethyl-5,6,7,8-tetrahydrobenzo[1,2,4]triazin-3-yl)-[2,2′] bipyridine (CyMe4-BTBP)/ – *N*,*N*′dimethyl-*N*,*N*′dioctyl-hexylethoxymalonamide (DMDOHEMA) – octanol	Am,Cm from Ln	ref. 13

OK = odourless kerosene (diluent), TBP = tri-*n*-butyl phosphate, FP = fission products, Ln = lanthanides, An = actinides, TRU = transuranic elements, HLW = high level waste.

with the former dominating the radioactive content of the wastes. Internationally, deep geological disposal, usually preceded by some decades of cooling to limit heat and radiation load on the host rock, is always assumed to be the management route for spent fuel. There is currently no operational disposal facility for spent fuel.

In a closed fuel cycle, as described above, there are many more options for the management of different streams. As a country which has operated a closed nuclear fuel cycle for over 50 years, the UK's approach to managing these different waste streams is fairly typical. Conventional Purex processes will produce uranium and plutonium product streams, and a liquid high level waste stream, which in current thinking will be vitrified for deep geological disposal. Concepts for disposal of vitrified high level waste are generally quite similar to those for spent fuel, because the heat and radiation loads are similar. However, as outlined above, removal of most of the actinide inventory would allow the hazard from vitrified high level waste to decrease faster than that from spent fuel.

Operations at all stages of either a closed or open fuel cycle will generate lower activity wastes associated with uranium mining, fuel fabrication, energy generation and spent fuel management. These are generally classified on the basis of their radioactive content (in the UK, classification is in decreasing order of radioactivity: intermediate level, low level and very low level wastes; see Chapter 6). In a closed fuel cycle, a wide variety of process wastes, for example ^{14}C- or ^{85}Kr-containing gases from fuel shearing and dissolution, or water from storage ponds, is also created. These are decontaminated where necessary (for example stripping of $^{14}CO_2$ from gases by precipitation as $BaCO_3$, or removal of ^{90}Sr and ^{137}Cs from aqueous effluents by ion exchange) prior to discharge to the environment under regulatory authorisation. The radioactive wastes from an open fuel cycle tend to be smaller in volume and less diverse than from a closed one. Obviously, the total activity is not changed.

4.1 The UK Waste Inventory

The UK has a complex and diverse inventory of radioactive wastes described in some detail by Defra/NDA[8] and the associated documents. Waste is classified primarily according to its radioactivity content. Low Level Waste (LLW) is waste with a radioactive content below defined levels (4 GBq $tonne^{-1}$ α-emitters, or 12 GBq $tonne^{-1}$ β- and β,γ-emitters). Intermediate Level Waste (ILW) exceeds the threshold activity for LLW but is not so radioactive that it requires active cooling. High level waste (HLW) is the intensely radioactive fission product stream derived from fuel reprocessing and requires constant cooling due to the decay heat it generates.

In addition, the UK has materials which may be declared wastes in the future, and will then need to be managed accordingly. These include separated plutonium, where a small proportion is unfit for reuse in fuel and will have to be disposed as waste. Government is presently considering management options

Table 5 The UK Inventory of Radioactive Wastes and Potential Wastes. Amounts and activities are sums of actual and committed quantities. Data compiled from CoRWM,[15] Defra/NDA[8,16] and Baldwin *et al.*[5]

Material	*Volume/Mass*	*Activity (Bq)*
LLW	3200000 m^3	5.6×10^{14}
LLW not suitable for disposal[a]	<30000 m^3	$<1.0 \times 10^{14}$
ILW	240000 m^3	4.1×10^{18}
HLW	1730 m^3	7.7×10^{19}
Plutonium	102 tHM	4.0×10^{18}
Uranium	160000 tHM	3.0×10^{15}
Spent Fuel	7700 tHM[b]	3.3×10^{19}

[a]A proportion of LLW is unsuitable for disposal in the Low Level Waste Repository and will be disposed with ILW.
[b]Spent fuel comprises 5500 tHM of AGR fuel and 1200 tHM of Sizewell B PWR fuel.

for the rest of the plutonium stockpile, and one option is to declare it all as waste. The UK also holds large quantities of uranium (depleted, natural and reprocessed) in various forms which may be declared as waste. Finally, it is planned to dispose of a proportion of UK spent fuel as waste without reprocessing. Current plans also assume that any spent fuel from new build reactors will eventually be disposed as waste. Collectively, ILW, HLW, plutonium, uranium and spent fuel are often referred to as "Higher Activity Wastes" (HAW). The UK inventory of wastes and potential waste materials, excluding new build waste, is summarized in Table 5, and options for their management are discussed in Chapter 5.

5 Impact of the "Global Nuclear Renaissance"[i]

5.1 Growth in Demand

As more countries pursue industrial and technological development, with the associated increases in energy demand, and it becomes increasingly unacceptable to use carbon-based fuels in large quantities, alternative sources of energy are required. Nuclear power has been out of favour for a generation in North America and much of western Europe, partly as a result of the Three Mile Island (1979) and Chernobyl (1986) accidents, but is now being re-evaluated and new reactors are proposed in the UK. Current UK proposals are for these new reactors to operate an open fuel cycle. The impact of Fukushima on these proposals is not yet clear. In Germany, there is public and political opposition to new nuclear build, and proposed life extensions for existing reactors have been thrown into doubt following the Fukushima accident. Sweden's government changed a 30-year old policy of phasing out nuclear

[i]The majority of the information in this section is taken from the World Nuclear Association (http://www.world-nuclear.org/info/).

energy in 2010, and the existing reactors are being uprated and their lives extended. Switzerland plans to retain nuclear as part of its energy mix and build replacements for existing reactors as they reach end of life. Italy, having closed its last reactors in 1990, has imported large amounts of nuclear electricity from France and had started to consider new build, though these plans are now delayed by the Fukushima accident. France, by contrast, has consistently developed nuclear electricity generation. It currently operates 58 reactors, providing 75% of its electricity, and intends to replace these from 2020, which implies construction of about one new reactor per year for 40 years. France is also actively pursuing more advanced nuclear technologies (fast reactors and Generation IV systems). In the USA, after 30 years in which no new reactors were built, 16 applications to build a total of 24 new reactors have been made since 2007. However, the existing 104 reactors, many of which have had life extensions, continue to provide about 20% of US electricity.

In other regions, notably Asia, some nations such as Japan and South Korea have pursued nuclear energy over many years. China currently has 12 reactors, with 24 under construction, and further plans for large scale expansion. Japan aims to expand nuclear generation from 30% of electricity production today to 40% by 2017, and also sees nuclear power as an important part in a 90% reduction of carbon emissions by 2100. South Korea is now positioned to export reactor technology. In the Middle East, several nations are exploring nuclear generation. Iran is in the process of fuelling a Russian-built reactor, while the United Arab Emirates have ordered four South Korean reactors.

Worldwide, there are currently 440 reactors operating, and projections of future growth in nuclear capacity, while they are obviously very uncertain, range from a 2–4 fold expansion by 2030, 3–10 fold by 2060 and 6–30 fold by 2100.

5.2 Implications for the Fuel Cycle

A modern uranium-fuelled PWR contains 400–600 tonnes of uranium in fuel, of which about a third is replaced on average each year. To support a global reactor fleet of 440, annual fuel production, at a typical enrichment of 3.5% ^{235}U, will need to be about 8–10000 tonnes, implying uranium production of 60–70000 tonnes yr^{-1}. Currently, about 25% of this need is being met by "blending down" surplus high enriched uranium from military programmes, but this will not continue over the long term.

Uranium is not a particularly rare element (crustal abundance 2.8 ppm) and in 2007, global uranium reserves were estimated at 5.5×10^6 tonnes.[ii] There is a

[ii] "Resources" are the total quantity estimated to be available; "reserves" are that proportion of the resources which can be extracted economically. Clearly, the balance between resources and reserves varies with the uranium price. As price increases, more resources can be exploited economically and therefore become reserves. Also, a higher price prompts exploration and the identification of additional resources and reserves. Resources and reserves can also change as a result of better definition of deposits, or successful exploration. The data presented here are based on a uranium price of \$130 kg^{-1}. The uranium price in October 2010 was \$138 kg^{-1}.

comparable quantity of uranium in seawater (4.5×10^6 tonnes; mean seawater concentration 3 ppb) but it is presently difficult to envisage a cost effective, large scale process for extracting it. The current position is therefore that global reserves would provide about 100 years' fuel at current consumption rates, but expansion even at the lower end of the scale projected will reduce this to a few decades, comparable to the lifetime of a modern reactor. However, there is a complex relationship between demand, price, exploration activity and size of reserves, so it is difficult to draw firm conclusions about the long term availability of uranium.

Even so, given the very long lead times associated with nuclear technology, a debate about alternatives has to be conducted at some point over the next decades. An open uranium fuel cycle is arguably wasteful and appears not to be sustainable over more than a century or two. A closed fuel cycle, particularly if combined with fast reactors, offers a vast increase in energy availability, but at the cost of industrial scale fuel reprocessing, which is a difficult and costly technology, and the large scale creation of plutonium or other fissile materials, which brings with it major ethical and security issues. Other fission technologies, such as thorium-fuelled reactors, would raise similar technical and ethical questions. Probably the most far-reaching question is therefore the role we see for nuclear fission? Is it a stopgap, lasting a few decades and bridging from a fossil fuel era to a renewable- or fusion-powered era, or is it a resource we will need to exploit over centuries? The answer to this question has substantial implications for the fuel cycle(s) we choose to develop, and the associated environmental impacts.

6 Conclusions

Nuclear fission potentially offers the prospect of very substantial amounts of energy from a low carbon source. However, all steps in the nuclear fuel cycle create wastes and have potentially major environmental impacts. The open fuel cycle creates smaller waste volumes and, at first sight wastes which are easier to manage, than a closed fuel cycle, but involves the disposal as waste of large quantities of potentially reusable material. The technology required for closed fuel cycles, for fast reactors, or for partitioning of long-lived waste components is particularly demanding and fast reactors, separations beyond Purex, and partitioning and transmutation in particular are far from mature. Likewise, many aspects of the conditioning and disposal of higher activity wastes remain challenging. In addition, the proliferation risks associated with the widespread production and use of fissile materials must be addressed. While the demand for nuclear energy appears to be growing substantially at present and is expected to do so in future, this raises complex questions for the long term, to which there are currently few clear answers.

Acknowledgements

This work is funded by the RCUK Energy Programme through its support of the MBase consortium.

References

1. A. Abdelouas, Uranium mill tailings: geochemistry, mineralogy, and environmental impact, *Elements*, 2006, **2**, 335–341.
2. NDA, *Plutonium Options*, Nuclear Decommissioning Authority, UK, 2008.
3. WNA, 2010, http://www.world-nuclear.org/info/inf29.html (accessed 4th October 2010).
4. M. J. Carson, *Reactor Grade Plutonium's Explosive Properties NPT95*, Nuclear Control Institute, Washington DC, USA, 1990.
5. T. Baldwin, N. Chapman and F. Neall, *Geological Disposal Options for High-Level Waste and Spent Fuel*, Report for the UK Nuclear Decommissioning Authority, UK, Jan 2008.
6. A. Hedin, *Spent Nuclear Fuel – How Dangerous is it*? SKB Technical Report 97–13, SKB, Stockholm, Sweden, 1997.
7. M. S. Gerber, *The Plutonium Production Story at the Hanford Site: Processes and Facilities History*, Report WHC MR 0521, Westinghouse Hanford Co, Richland, WA, USA, 1996.
8. Defra/NDA, *The 2007 UK Radioactive Waste Inventory*, Defra/RAS/08.002 2008a NDA/RWMD/004, 2008.
9. E. P. Horwitz and W. W. Schulz, The TRUEX process: a vital tool for disposal of US defense nuclear waste, in *New Separation Chemistry for Radioactive Waste and Other Specific Applications Rome*, CONF-900579-2, Italy, May 16–18, 1990.
10. C. Madic, P. Blanc, N. Condamines, P. Baron, L. Berthon, C. Nicol, C. Pozo, M. Lecomte, M. Philippe, M. Masson, C. Hequet and M. J. Hudson, Actinide partitioning from HLLW using the DIAMEX process, *Proceedings of the Fourth International Conference on Nuclear Fuel Reprocessing and Waste Management*, RECOD'94, London, 1994.
11. R. S. Herbst, J. D. Law, T. A. Todd, V. N. Romanovskiy, V. A. Babain, V. M. Esimantovskiy, I. V. Smirnov and B. N. Zaitsev, Universal solvent extraction (UNEX) flowsheet testing for the removal of cesium, strontium and actinide elements from radioactive, acidic, dissolved calcine waste, *Solvent Extract. Ion Exchange*, 2002, **20**, 429–445.
12. J. Brown, M. J. Carrott, O. D. Fox, C. J. Maher, C. Mason, F. McLachlan, M. J. Sarsfield, R. J. Taylor and D. A. Woodhead, Screening of TODGA/TBP/OK solvent mixtures for the grouped extraction of actinides, *IOP Conf. Ser.: Mater. Sci. Eng.*, 2010, **9**(1), (doi:10.1088/1757-899X/9/1/012075).

13. D. Magnusson, B. Christiansen, M. R. S. Foreman, A. Geist, J.-P. Glatz, R. Malmbeck, G. Modolo, D. Serrano-Purroy and C. Sorel, Demonstration of a SANEX Process in centrifugal contactors using the CyMe4-BTBP molecule on a genuine fuel solution, *Solvent Extract. Ion Exchange*, 2009, **27**, 97–106.
14. NEA, *Actinide Separation Chemistry in Nuclear Waste Streams and Materials*, NEA/NSC/DOC(97)19, Nuclear Energy Agency, Paris, France, 1997.
15. CoRWM, *Managing Our Radioactive Waste Safely – CoRWM's Recommendations to Government*, CoRWM Document 700, 2006.
16. Defra/NDA, *Radioactive Materials Not Reported in the 2007 UK Radioactive Waste Inventory*, DEFRA/RAS/08.005 2008b NDA/RWMD/007, 2008.

Nuclear Accidents

J. T. SMITH

ABSTRACT

In the wake of the Fukushima accident, this chapter provides a summary and comparison of the four previous major accidents in the history of exploitation of nuclear power for military and civilian purposes: Windscale, Kyshtym, Three-Mile Island (TMI) and Chernobyl. The events leading to each accident, and their consequences to environmental and human health, are summarised. The earlier accidents at Windscale (UK) and Kyshtym (former Soviet Union) could be attributed in large part to the pressures to produce plutonium for atomic weapons programmes during the early years of the Cold War. This led to nuclear facilities being built with insufficient emphasis on design safety and, in some cases, lack of full understanding of the processes involved. The latter accidents at TMI (USA) and Chernobyl (former Soviet Union) were also in part caused by design and equipment failures, but operator errors (caused by poor training, insufficient or unclear information and a failure in safety culture) made a key contribution. In terms of environmental and human health impacts, the Kyshtym and Chernobyl accidents were of much greater significance than those at Windscale and TMI. Both Kyshtym and Chernobyl caused mass permanent evacuation and significant long-term environmental contamination. As demonstrated at TMI, even where radiation doses to the public are very low, psychological and social consequences of nuclear accidents can be serious. Concerning impacts of nuclear accidents on ecosystem health, there is no clear evidence that even the Kyshtym and Chernobyl accidents have caused significant damage in the long term. However, studies of the effects of radiation damage in these contaminated environments have been confounded by the largely positive impact evacuation of the human population has had on the ecosystem.

Issues in Environmental Science and Technology, 32
Nuclear Power and the Environment
Edited by R.E. Hester and R.M. Harrison

Published by the Royal Society of Chemistry, www.rsc.org

1 Introduction

Prior to the 2011 Fukushima accident, there had been four accidents of major importance in terms of their actual or potential consequences for the environment and human health. The two earliest accidents, at the Windscale site in the UK and at Kyshtym in the former Soviet Union, were at facilities which were part of the Cold War drive to produce materials for nuclear weapons production. The latter two, at Three-Mile Island (TMI) in the USA and at Chernobyl, in the former Soviet Union, were at civilian nuclear power plants (see Table 1). Much has previously been written about these four accidents, particularly the Chernobyl accident, and this chapter aims to give a brief summary and overview of this literature.

In addition to these major accidents, there have previously been significant releases of radioactivity to the environment during development of nuclear weapons and from "routine" operations at nuclear facilities. Some key past releases of radioactivity to the environment are summarised in Table 2. It should be noted that long-term environmental contamination from the Hiroshima and Nagasaki bombs was not significant; the high radiation exposures to the population of these cities came primarily from radiations during or shortly after the explosions. During the Cold War, atmospheric (*i.e.* above ground) testing of nuclear weapons caused fallout of relatively low level radioactivity (particularly ^{90}Sr, ^{137}Cs and ^{14}C) globally, mainly in the Northern Hemisphere. Several hundred atmospheric nuclear weapons tests were carried out by the USA, USSR and the UK until a test ban treaty was signed in 1963. Limited atmospheric nuclear weapons tests were carried out by France and China in the early 1970s.

2 The 1957 Windscale Fire

2.1 Events Leading to the Accident

In the early hours of the morning on the 10th of October 1957 a reactor (Number 1 of the two Windscale "Piles"), developed to produce plutonium and tritium for the UK's atomic bomb programme, caught fire. The fire occurred during a procedure to release so-called "Wigner" energy from the reactor core.

Table 1 Summary of the four nuclear accidents considered here.

Accident	*Year*	*Sector*	*INES Category*[a]
Windscale	1957	Military	5
Kyshtym	1957	Military	6
Three-Mile Island	1979	Civilian	5
Chernobyl	1986	Civilian	7

[a]International Nuclear Events Scale: Category 5 = accident with wider consequences; Category 6 = serious accident; Category 7 = major accident.

Table 2 Summary of previous major releases of radioactive material to the environment. Note that the summary is not comprehensive and only data for three radionuclides are presented. These release data should not be interpreted in terms of significance of the releases to environmental or human health; the impact of radionuclide releases is not solely determined by the amount of radioactivity released. (Adapted from Smith & Beresford).[48]

		Release of some key radionuclides to the environment (PBq)		
Release event	*Area*[b]	*^{137}Cs*	*^{90}Sr*	*^{131}I*
Chernobyl, 1986 (ref. 39)	Significant part of Europe	85	10	1760
Hiroshima atomic bomb, 1945[a] (ref. 69)	Few km radius around epicentre	0.1	0.085	52
Atmospheric nuclear weapons testing, 1952–1981** (ref. 70)	Global, primarily Northern Hemisphere	949	578	**
US atmospheric weapons tests, Nevada Test Site**,[71]	US states, particularly Nevada	**	**	5550 released to atmosphere, 1390 deposited to ground
Three Mile Island, USA[31,27]	No significant environmental contamination	–	–	4.81×10^{-4}
Mayak, discharges to the Techa River, 1949–56 (ref. 18)	Techa and Ob ivers	13	12	n.d.
Kyshtym accident, 1957 (ref. 18)	Approx. 300 km × 50 km area of Siberia	0.027	4	–
Waste discharges from Sellafield, 1964–92 (ref. 72)	Irish Sea	41	6	n.d.
Windscale accident, 1957 (ref. 1)	518 km^2 area of Northern England	0.18	7.5×10^{-4}	1.8

[a]Note that the radiation health effects of the Hiroshima and Nagasaki bombs resulted primarily from gamma and neutron radiation from the initial explosion. Radioactive fallout to the environment (detailed here for Hiroshima) was minor in comparison.
[b] Indicative area only: the contaminated area depends on how you define "contaminated".
** ^{131}I data is given for the US atmospheric weapons tests only: ^{137}Cs and ^{90}Sr data are global totals for the period 1952–1981.

Wigner energy is chemical potential energy stored in the lattice structure of the graphite moderator during operation of this type of nuclear reactor. The operators at Windscale routinely released this stored energy (to prevent an uncontrolled release) by an annealing process in which the core temperature was temporarily raised. This procedure normally released the Wigner energy in the graphite resulting in a temporary heating, then cooling of the core. On the day of the accident, however, the annealing process caused the temperature of some parts of the core to rise substantially. This, possibly coupled with rupture of a fuel element, caused the reactor to set on fire. After failed attempts to remove the overheated fuel elements and to put out the fire by carbon dioxide, the reactor was flooded with water on the following day (11th of October). The operators believed at the time that use of water carried the risk of a hydrogen explosion as it contacted red-hot metal, but it was felt that the risk of breach of containment by the burning reactor core was greater. By the evening of the 11th, the fire was fully extinguished.

The fire resulted in the release of "some of the fission products and activation products contained in a few percent of the core".[1] The releases from the Windscale fire have recently been re-evaluated,[1] giving estimates for a wide range of radionuclides including 1.8 PBq of ^{131}I; 0.18 PBq of ^{137}Cs and 0.042 PBq of ^{210}Po. As with TMI and Chernobyl, large quantities of noble gases (including 26 PBq of ^{133}Xe) were released, though these were less radiologically significant.

2.2 *Environmental Contamination*

Radioactivity, most importantly ^{131}I, ^{137}Cs and ^{210}Po, was carried to the East by prevailing westerly winds, though the wind speed and direction varied during the course of the accident.[1,2] Deposition to the ground across Northern England "was dominated by ^{131}I (half-life of 8.04 days) with deposits above 4 kBq m^{-2} extending about 75 km east northeast and 140 km south southeast of the site, covering an area of about 12 000 km^{2}" (Jones[2] citing Chamberlain).[3] Iodine-131 was measured across the North Sea in Holland and Belgium, though concentrations were much lower than in England. Figure 1 shows a map of the iodine release. A recent re-analysis of air monitoring data in Norway[4] showed that the radioactive plume reached Norway on the 15th and 16th October. These authors noted that maximum observed deposition was comparable to the level of deposition from atmospheric nuclear weapons testing in 1958. It has been estimated by Garland and Wakeford[1] that no more than 10% of the total ^{131}I passed across the East coast of England to the North Sea and Europe.

As the accident progressed, in the early hours of October 11th, the regional police chief constable was notified.[5,6] A review of the accident[6] concluded that, after the uncertainty and confusion of the initial incident, the aftermath was handled well: "community warnings and communications were handled efficiently and promptly, environmental survey teams and equipment were assembled and dispatched promptly, and there was an atmosphere of quiet professionalism".

Assessment of external radiation doses showed that these were not high enough to require evacuation (Jackson and Jones[7] citing Dunster *et al.*[8]), but

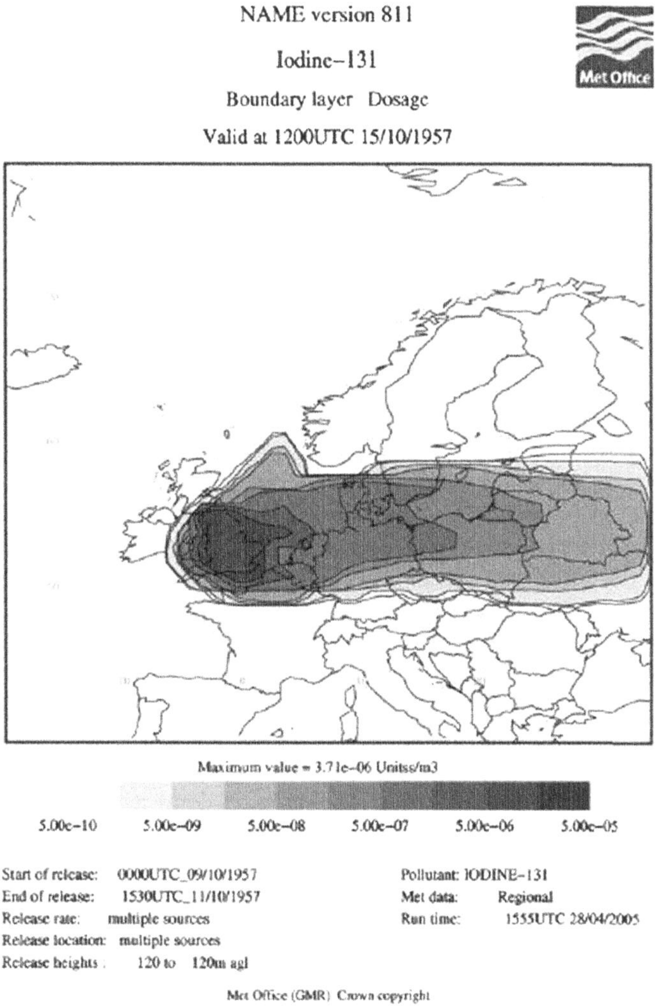

Figure 1 Time integrated concentrations of ^{131}I in air following the Windscale accident up to 12.00 on the 15th of October 1957. (Reprinted from Johnson *et al.*[66] with kind permission of Elsevier).

high levels of radioactivity (principally ^{131}I) in milk implied potentially significant ingestion doses. The majority of iodine ingested by the body is accumulated in the thyroid, so radioactive iodine intake results in a risk of thyroid cancer. At the time, no intervention level for ^{131}I in milk had been set, so, in the words of Jones[2] "hasty, but effective, consultations and calculations" determined a maximum permissible level of 0.1 μCi l^{-1} (3700 Bq l^{-1}). A sampling campaign was set up to determine levels of radioactivity in milk across a large area of the UK. In the local area on the 11th and 12th of October, observed ^{131}I activity concentrations in milk reached 30 000 Bq l^{-1}, but rapidly declined over the following weeks. Similar maximum activity concentrations were observed in other foodstuffs.[7] Levels in public drinking water sources were not expected to be high.

Approximately three million litres of milk were discarded over an area of around 500 km^2 (ref. 7). Restrictions were finally lifted on the 23rd of November, approximately six weeks after the accident.[2] It is probable that, at present day intervention levels, the area in which milk consumption was banned would have been much greater[7] and temporary precautionary bans on foodstuffs, including meat and milk, would also have been implemented as a consequence of radiocaesium contamination.[7,9]

2.3 Radiation Exposures and Health Impacts

The ban on milk consumption significantly reduced the radiological impact of the accident on the population. Jackson and Jones[7] estimated that the ban averted 75% of the ingestion dose to children, and a higher proportion of the ingestion dose to adults. The maximum dose to the thyroid of children in the local area was 160 mSv, with average doses being in the range 10–100 mSv.[7] As a result of the milk ban, the main dose pathway was inhalation,[10] though in children there was also a significant contribution from ingestion.[7] The collective effective dose equivalent from external radiation, inhalation and ingestion was approximately 1900 person-Sv in the UK and 100 person-Sv in the rest of Northern Europe.[10] Estimates imply that about 50% of the collective effective dose equivalent was due to inhalation, the remainder being mainly due to ingestion of milk and other foods (about 35%), the rest being attributed to external radiation from the cloud and ground deposits. Clarke[11] estimated approximately 100 fatal and 100 non-fatal cancers in the UK population (over a 40–50 year period) resulting from the Windscale fire release. Polonium-210 was expected to give rise to the majority of fatal cancers. Thyroid cancer (from ^{131}I), being in most cases successfully treatable, would be expected to form the majority of non-fatal cancers.

The median dose to 466 workers involved in fighting the fire and in clean-up work was 3.52 mSv with the maximum individual dose being recorded as 43.9 mSv (determined from monthly dose monitoring records for October 1957).[12] It was reported[12] that the collective dose to these workers was 2.33 person-Sv for October 1957, which was approximately double the average monthly collective dose for 1957. As might be expected from the low median individual and collective doses, a study[12] of mortality and the number of registered cancers during the period 1957–97 was "unable to detect any effect of the 1957 fire upon the mortality and cancer morbidity experience of those workers involved in it".

2.4 Social and Psychological Consequences

There is relatively little information available on social and psychological consequences of the Windscale accident. This likely reflects a lower level of awareness, both in the public and in scientists and decision makers, of these issues at the time. A great deal of secrecy surrounded the Windscale plant and accusations have been made of a "cover-up" of the accident consequences by the operators and authorities.[13] The fact that Wolff[5] noted that the contaminated milk from the accident "...could have been used for manufacturing purposes or the feeding of livestock but, because of public apprehension, it was decided not to salvage the milk" suggests significant public concern over the contamination, and official awareness of that concern. An article in *Scientific American* a few months after the accident stated that "The accident produced something approaching panic among the local population".[14]

3 The Kyshtym Explosion

3.1 Events Leading to the Accident

The explosion, on the 29th of September 1957, of a high-level waste tank at the Mayak plutonium production and reprocessing facility in Siberia was, until Chernobyl, the World's most severe nuclear accident. Since the Mayak site and associated town Ozyorsk were secret and as such were not found on Soviet maps, the accident was named after the next nearest town of Kyshtym. The first official acknowledgement of the accident by the Soviet Union was in June 1989 (ref. 15). The stored wastes had been treated to remove radiocaesium, but contained various other fission products. The concrete storage tank had a volume of 300 m^3 and contained, at the time of the explosion, 70–80 tonnes of liquid waste.[16] The cause of the accident was the failure of the tank's cooling system, which caused the waste to heat up. The high temperature and the evaporation of water from the tank resulted in a chemical explosion of nitrate and acetate compounds in the waste. It is believed that all of the 740 PBq of waste was released from the tank, 90% of which remained within the site, the remaining 10% being dispersed to the environment in a plume of aerosols which reached a height of one kilometre,[16] as illustrated in Figure 2.

3.2 Environmental Contamination

The stored wastes had been treated to remove radiocaesium, so were largely free of this radionuclide, but contained various other fission products produced in the nuclear reactors at Mayak. Prior to the accident, the wastes had been stored for approximately one year[17] so they contained no very short-lived radioisotopes (for example, the radiologically important ^{131}I had decayed away by the time of the accident). According to a 1989 report to the International Atomic Energy Agency,[18] the accident released approximately 74 PBq of beta- and gamma-emitting radionuclides to the wider environment (10% of the total release), comprising mainly (94.6%) of radionuclides of half-life of one year or less (^{95}Zr, ^{95}Nb, ^{106}Ru, ^{106}Rh, ^{144}Ce and ^{144}Pr) as well as the relatively long-lived (half-life of 28.8 yr) ^{90}Sr (5.4%; 4 PBq). The decline in activity present in

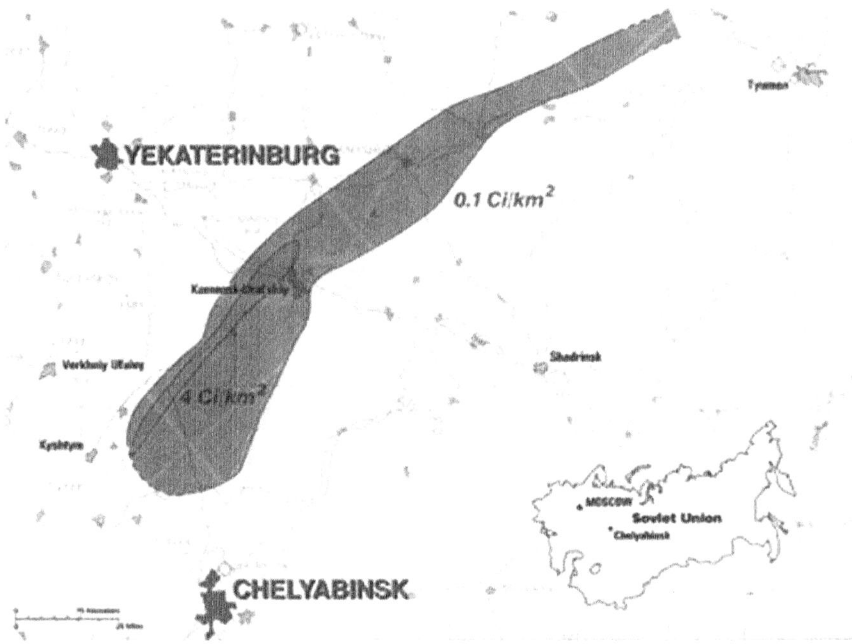

Figure 2 Map of ^{90}Sr deposited following the Kyshtym accident, named the "East Urals Radioactive Trace" from Dicus.[67]

the environment (due to radioactive decay) is illustrated in Figure 3, showing a relatively rapid decline in the years after the accident, though the decline was slower than that at Chernobyl because of the absence of very short-lived fission products in the stored waste.

The extent of ground contamination was defined[18] as land which had ^{90}Sr contamination density greater than 3700 Bq m^{-2}. By this definition, the contamination trace (known as the East Urals Radioactive Trace; EURT) extended 300 km from the Mayak site in a relatively narrow path covering 15 000–23 000 km^2 (ref. 17,18). The area in which contamination was greater than 74000 Bq m^{-2} of ^{90}Sr was approximately 105 km long and 8–9 km wide, covering 1000 km^2 (ref. 18; see Table 3). Within ten days, six hundred people had been evacuated from the areas of highest contamination density, and over a period of two years, all 10 000 people living within the $>$74 kBq m^{-2} area were evacuated. Over the following years, evacuated lands were gradually returned to agricultural production and, by 1978, the majority had been reclaimed.[18]

3.3 Radiation Exposures and their Environmental and Health Impacts

Initially, radiation doses to people were primarily from external gamma radiation, but after nine months, ingestion of contaminated products became

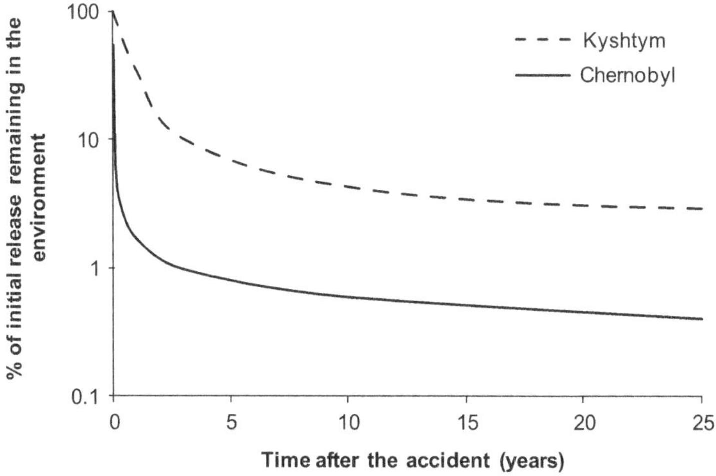

Figure 3 Declines in total radioactivity in the environment during the years after the Kyshtym and Chernobyl accidents. (From data in Nikipelov *et al.*[18] and Smith and Beresford).[48]

Table 3 Areas contaminated by the Kyshtym accident, and size of affected population.[17,18]

Sr-90 contamination density ($Bq\ m^{-2}$)	*Area affected (km^2)*	*Population*
3.7×10^3	23 000	270 000
7.4×10^4	1000	10 000
3.7×10^6	120	2100

the most important dose pathway, as relatively short-lived beta/gamma emitters decayed away and (the pure-beta emitting) ^{90}Sr began to dominate.[17] In the first years after the accident, ingestion of ^{90}Sr was primarily due to consumption of contaminated bread and milk as well as water from local reservoirs.[17] It was also reported[17] that, eight years after the accident, foodstuffs contributed to ingestion doses in the following order: milk (50%); vegetables (15%), potatoes, 12%, eggs (8%), meat (7%) and bread (4%). Note that these values are likely to be strongly influenced by countermeasures to control and reduce activity concentrations in foodstuffs. Since strontium has a similar uptake chemistry to calcium (both being alkaline earth metals), products rich in calcium tended to have high ^{90}Sr concentrations. In the human body, strontium is incorporated into bone and teeth, so doses to the bone marrow can be high.

In 1987, thirty years after the accident, it was reported,[17,18] that the ^{90}Sr content in farm products had declined significantly compared to one year after the accident. Following atmospheric nuclear weapons testing, radiostrontium activity concentrations in plants were observed typically to decline with an effective half life in the range 8–14 years[19,20] due to physical decay, loss of ^{90}Sr from the upper soil layers and changes in soil adsorption. At Kyshtym, in addition to these natural processes, Alexakhin *et al.*[17] report an extensive system of countermeasures which played an important role in reducing ^{90}Sr activity concentrations in foodstuffs. These countermeasures included deep-ploughing of contaminated land to reduce radioactivity in the rooting layer of plants, liming of acid soils and addition of calcium to the feed of any calcium-deficient farm animals.

The radiation dose to the population varied greatly, being dependent to a large extent on the contamination density. Average doses (received before evacuation) were 520 mSv to the 600 people living within the 18.5 MBq m^{-2} zone who were evacuated within 20 days. The 3100 people living in areas of contamination density of 0.12–0.33 MBq m^{-2} received 23 mSv in the 670 days before their evacuation was completed.[18] In areas with a ^{90}Sr contamination density of 0.037 MBq m^{-2} which were not evacuated, Alexakhin *et al.*[17] reported an effective equivalent dose over a 30 year period of 12 mSv. Relatively recently, it was reported[21] that "At present, doses to members of the public living in the area of the East-Ural radioactive trace are significantly lower than 1 mSv/year".

The overall average radiation dose to the 270 000 people who were living within the EURT at the time of the accident was reported to be 7 mSv, with a collective dose of approximately 2000 person-Sv.[22] Assuming a fatal cancer risk factor of 0.05 Sv^{-1}, this implies an average individual risk of *ca.* 1 in 3000 and an expected 100 fatal cancers in the affected population.

The doses and dose rates to the population were therefore likely to give rise to a significant increase in fatal cancers, though owing to the relatively small number of people exposed to the highest doses and incomplete follow-up, this may not have been possible to observe in the epidemiological evidence. A summary of this evidence[18] suggests no observable increases in deterministic symptoms (radiation sickness) or serious illness, though a reduction in leukocyte blood count was observed in the exposed population. The radiation doses to the exposed population at Kyshtym (at least, the reported average dose to the most exposed people) was significantly below lethal levels, though it is possible that some deterministic effects would have been seen in some of the most exposed of this group.

Kossenko[22] summarised the epidemiological data as follows:

> *"The cancer mortality rate was estimated for 7800 people exposed due to the Kyshtym accident and moved from the EURT territory, and for 8000 people left to reside in that area. In the four groups studied, with doses averaging from 0.54 to 0.006 Gy [presumably approximately equal to 0.54 to 0.006 Sv for this beta/gamma radiation], no increase in the incidence of lethal leukaemia and solid cancers was found. In addition, the investigation did not reveal disorders of the reproductive function in the exposed people (according to birth rate data) nor any increase in the incidence of developmental defects among the offspring of people exposed to radiation due to the Kyshtym accident".*

Workers at the Kyshtym site during and after the accident received higher doses than the population. According to Kruglov (2002), 5000 workers at the site at the time of the accident received doses up to 1 Sv and 30,000 clean-up workers received doses above 250 mSv during 1957–59.[74]

There were a number of long-term studies on the effects of Kyshtym on biota. In the most contaminated "head" of the EURT, acute radiation syndrome was observed in farm animals, in many cases leading to death. Within 12 days of the release, estimated whole-body doses to some farm animals reached 2.9 Gy and doses to the gastrointestinal tract (GIT) were in the region 20–50 Gy.[17]

Longer term studies also showed evidence of effects on organisms, though this evidence is not always conclusive and radiation exposures to animals were not always estimated, making interpretation difficult. A study on dandelion seeds in 1991 and 1993 (ref. 23) showed some increase in chromosome aberrations in contaminated compared to control areas, though this was not consistently observed. Germination and viability did not differ between contaminated and control areas.[23] In studies conducted from 1961–1991, Krivolutsky[24] observed that the number of species of soil-dwelling insects was lower by a factor two in a contaminated plot of birch forest compared to two control plots, but no significant difference was found between a contaminated agricultural system compared to a control one.

In aquatic systems, dose rates to benthic (bottom-dwelling) fish in the highly contaminated Uruskul' and Berdyanish lakes were up to 0.1 Gy d^{-1} during the first few months after the accident.[25] There were reported[25] "disturbances in the reproductive process of carp and goldfish... [which] manifested itself predominantly in the first few years following the accident". No effect on the number of fish species was found, attributed by the authors[25] to "the fact that the radiation effect was compensated for by a ban on commercial catches of fish".

3.4 Social and Psychological Impacts

There appears to be little information (at least in the English language) in the scientific literature on the social and psychological impacts of Kyshtym on the population. A report by Collins[15] highlights the delays in evacuation of many contaminated areas and the lack of information, and humanitarian and medical aid given to exposed populations. This contributed, unsurprisingly, to strong anti-nuclear feelings amongst the Kyshtym population termed by Collins[15] a "psychological animosity" towards nuclear power.

4 Three-Mile Island

4.1 Events Leading to the Accident

The accident at reactor number two at the Three-Mile Island (TMI) nuclear power station in Pennsylvania, USA, occurred on the early morning of the 28th of March 1979. Whilst leading to relatively little off-site environmental contamination, the accident caused a partial core meltdown and was considered to be the most serious accident at a civilian nuclear power station prior to Chernobyl.

The accident was initiated by a failure in the secondary (non-nuclear) electricity generating part of the plant which caused a pressure increase in the cooling system of the primary (nuclear) part of the plant. The pressure was released by a valve but, once released, the valve failed to close, causing loss of cooling water.[26] Had proper action been taken at this stage, this would have been a relatively minor incident, but subsequent operator errors, caused by poor information and training[27] led to a major reactor failure. There was no indication to the operators that the pressure release valve had failed to close, nor was there a clear indication of the level of the coolant in the reactor core. With "more than 100" (ref. 27) alarms going off in the control room, the plant operators were not aware that coolant was being lost from the core. Thus, when an automatic safety system began to inject cooling water into the core, the operators overrode the system, drastically reducing the rate of injection of cooling water. The loss of coolant led to exposure and the resulting meltdown of a significant part of the reactor core.

4.2 Environmental Contamination

The Three Mile Island accident caused the release of radioactive gasses and volatile fission products from a 55 m high vent stack adjacent to the reactor building. Between 90 and 480 peta-Bequerels of radioactive noble gases were released but releases of the more radiotoxic ^{131}I (half-life of 8.1 days) were much lower, in the range 4.81×10^{-4} PBq.[27] Of the noble gases, it was estimated[28] from the core inventory at time of shutdown, that "1% was ^{88}Kr, 95% was ^{133}Xe and 4% was ^{135}Xe". All of these released radionuclides emit beta/gamma radiation.

The release of radioactive noble gases led to radiation exposures to local residents from external gamma and, to a lesser extent, beta-radiation:[29] inhalation exposures were negligible.[28] The noble gases did not present a risk from ingestion of contaminated food products and were quickly dispersed in the atmosphere. The accident at Three Mile Island did not result in significant contamination of the environment and food chain. According to Gerusky[30] "thousands of samples of milk, air, water, produce, soil, vegetation, fish, river sediment, and silt in the TMI vicinity were analyzed". The highest level of contamination in food products was found in a sample of goat milk 2 km from the site; the ^{131}I activity concentration[31] was only 1.5 Bq l^{-1}. Thus the internal radiation dose from ingestion of contaminated foodstuffs was very low; maximum doses to the thyroid of infants were in the order of tens of μSv. Whilst the environment was not significantly contaminated, large quantities of radioactively contaminated cooling water which escaped into the containment and auxiliary buildings were not fully decontaminated until 1993 (ref. 26). At the present time, the reactor is de-fueled and is currently in "monitored storage".[26]

4.3 Radiation Exposures and their Environmental and Health Impacts

Radiation exposures to residents local to Three Mile Island were difficult to accurately quantify since the direction and extent of the cloud of radioactive gasses was uncertain. According to Hatch *et al.*[29] "...the vent-stack monitor at the reactor went "off scale" and the thermoluminescent dosimeters [TLDs] located outside the plant provided incomplete coverage...". In a study covering the population living within 80 km of Three Mile Island, Miller *et al.*[28] used a Gaussian plume model to reconstruct radioactive noble gas concentrations at ground level and thus to estimate external doses. The study[28] found good agreement between the model and the (somewhat limited) TLD observations, though they noted the uncertainties in estimating plume travel, particularly around buildings. The estimated collective population dose was 15 person-Sieverts[28] distributed amongst 2.16 million people. Elsewhere,[29] the average dose (averaged over a much smaller, more local, population) was reported to be approximately 0.1 mSv with the maximum dose to any one person being 1 mSv. The average dose was many times lower than average annual radiation doses from natural background radiation.

These estimated radiation doses are very low. The enhanced risk of fatal cancer to the most exposed individual is estimated to be 1 in 20 000 and the 15 person-Sv collective dose would lead to, possibly, one additional fatal cancer in the exposed population of approximately two million people. It would therefore not be expected to be possible to detect any statistically significant cancer increase (or increase in other health outcomes such as birth defects) in the population affected by the Three Mile Island accident. This indeed was the conclusion of the 1979 Presidential Commission investigating the accident.[27] Since the accident, however, a number of studies have been carried out attempting to link radiation from Three Mile Island to cancer incidence and mortality. This was in part as a result of ongoing public concern over possible health impacts of the accident.

Due to the low and uncertain distribution of radiation exposures, it was not possible to carry out dose estimation and health follow-up on an individual basis. The epidemiological studies have, therefore, been limited to "ecological" studies of health effects in large population groups of different (estimated) average radiation exposure. Such studies are particularly prone to being confounded by other factors, such as smoking prevalence, which may cause different cancer rates in the exposed and unexposed groups, though attempts are usually made to account for these factors. In a study of 160 000 people living within a 10 mile radius of the power plant, Hatch *et al.*[29] divided the study cohort into 69 areas. Average doses to populations of these areas were categorised into four radiation dose groups based on a model for the plume of radioactive noble gases. Incidence data of a number of different cancers were collected for each area during four years before and seven years after the accident.

The Hatch *et al.*[29] study found no evidence of an increase in childhood cancers or leukaemia, both expected to be the most sensitive to radiation. No increase in other forms of cancer such as breast and thyroid (which is often associated with exposure to ^{131}I) was observed. An increase in non-Hodgkin's lymphoma (a cancer not believed to be associated with radiation exposure) was observed, though Hatch *et al.*[29] stated that "...the relation rests on a few cases and the result is therefore unstable. It would not be surprising, given multiple hypotheses, to find one such association simply by chance". An apparently statistically significant association was also found between lung cancer and radiation exposure, although the authors[29] stated that "confounding by smoking cannot be ruled out". It has further been argued[32] that none of the TMI epidemiological studies have adequately accounted for possibly differing levels of natural radon exposures between the dose groups, also making interpretation of any apparent excess in lung cancer rates difficult.

In a more recent study Wing *et al.*[33] re-interpreted the data of Hatch *et al.*[29] finding again an association of estimated dose with lung cancer and, in contradiction to the earlier study, a significant increase in both all-cancer and leukaemia incidence. However, a subsequent study[34] of 35 000 residents within a five-mile radius of the plant with a much longer 20-year (1979–1998), follow-up period "provided no consistent evidence that radioactivity released during the TMI accident ... has had a significant impact on the mortality experience of this cohort..." although they also noted that "...certain potential dose–response relationships cannot be definitively excluded".

4.4 Social and Psychological Impacts

Probably the most significant impact of the TMI accident was the stress caused to the local population by the potential risks to their health from radiation releases. Attention also focused on the potential for a catastrophic explosion of a bubble of hydrogen gas which had built up in the reactor during the days over which the accident unfolded. At 12.30 pm on Friday the 30th of March, two days after the accident began, state Governor Richard Thornburgh advised "pregnant women and school-age children to leave the area within a five-mile radius of the Three Mile Island facility until further notice".[35] This advice turned out to have been based on a meter reading showing high levels of radiation 130 feet above the plant which was, in the confusion, erroneously believed to be a ground-level reading in residential areas. Fears about an explosion of the hydrogen gas bubble within the reactor core surfaced on the Friday and continued until the Sunday, but again proved to be based on incorrect information on the state of the reactor which had been supplied to the US Nuclear Regulatory Commission.[27]

Following the Governor's advisory, more than 3500 pregnant women and children evacuated the area.[35] Uncertainty over radiation levels and risk of explosion led to an additional voluntary evacuation of a large number of other people (from both within and outside the five-mile area) during the days after the accident.[27,36] The Governor's advice to vulnerable groups to evacuate was lifted on the 9th of April.

The uncertainty surrounding the risks associated with the accident, the evacuation, and the ongoing perceived threat of the damaged reactor, all contributed to high levels of stress in the local population, both shortly after the accident[27] and in subsequent years. More than one year after the accident, Fleming *et al.*[37] and Baum *et al.*[38] observed elevated levels of a number of stress indices in the local population compared with control populations. Effects observed were "mild" but included a range of self-reported symptoms of emotional distress such as concentration problems, depression and anxiety.[38]

5 The Chernobyl Accident

5.1 Events Leading to the Accident

The explosion at Unit 4 of the Chernobyl nuclear power station was the worst nuclear accident in history. There are still some uncertainties regarding the exact causes and events leading to the accident, though the key factors are now known. The accident occurred during an experiment to test the behaviour of an electrical system which powered the station in the event of a failure of the main electricity supply. In order to conduct the experiment, the reactor power was reduced which (possibly due to a problem in the operation of the automatic control rods) led to the reactor being in an unstable state[39] and operating outside its design parameters.

At 01:23 on the morning of the 26th of April 1986, the experiment began, despite the fact that:

(i) The reactor power output was well below that required by the experimental procedure;
(ii) Certain reactor safety systems had been deliberately disabled in order to carry out the experiment; and
(iii) The number of control rods in the reactor was only half the minimum required for its safe operation.

Thirty seconds after the experiment began, the reactor power began to increase rapidly and ten seconds later the operators attempted a full emergency shut down by re-inserting the control rods. The reactor power was now increasing exponentially, leading to a failure in the pressurised cooling water system. Eight seconds later, the reactor exploded (an explosion of steam, not a nuclear explosion) scattering burning core debris over the surrounding area.

Over 100 firemen were called to the scene and they worked with plant personnel to put out many small fires in the reactor building and on the roofs of Unit 4 and the adjacent Unit 3 building. This work exposed the emergency workers to extremely high doses of radiation. During the days after the explosion, helicopters were used to dump thousands of tonnes of various materials onto the exposed reactor core. These materials included boron, lead, sand and clay to smother the fire, absorb radiation and reduce nuclear reactions in the molten core material. In total, 1800 helicopter flights were made[39] at great risk to the pilots. Despite the heroic efforts of firemen, helicopter pilots and many other emergency workers to put out the fire, the reactor continued to burn for ten days.

5.2 Environmental Contamination

The explosion and subsequent fire spread volatile radioisotopes over large areas of the former Soviet Union and parts of Western Europe (see Figure 4). The pattern of deposition of radioactivity was complex; outside the area surrounding the power plant, the highest deposition of radiocaesium isotopes

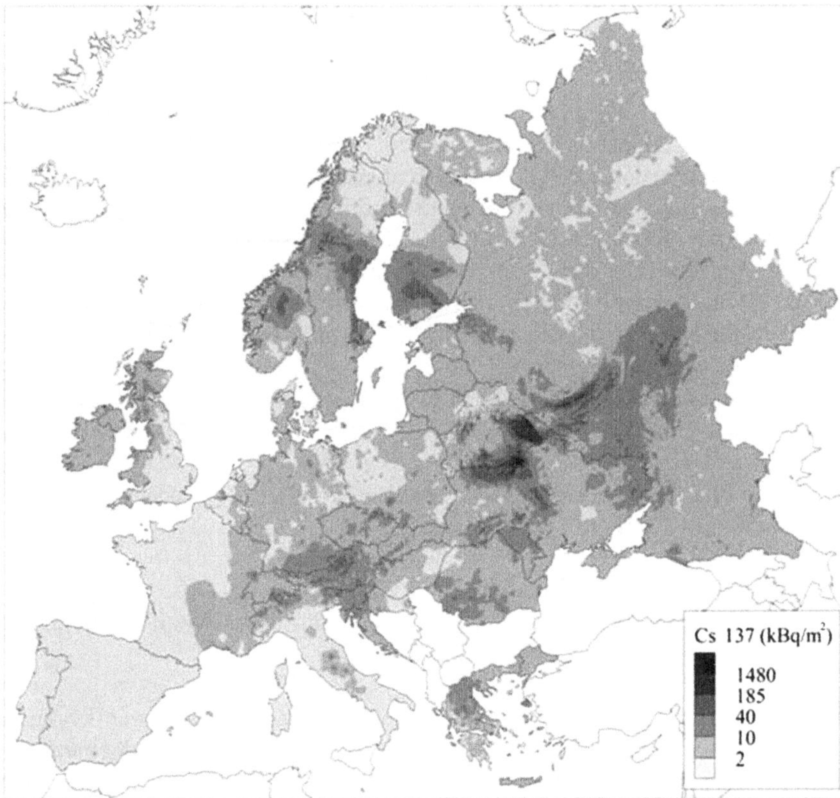

Figure 4 ^{137}Cs fallout in Europe from the Chernobyl accident. Cs-137 deposition was interpolated by Simon Wright (CEH Lancaster UK) from contours presented by De Cort.[68]

occurred in areas where rainfall intercepted the radioactive plume as it dispersed. Less volatile elements such as isotopes of strontium and plutonium were deposited principally within 30 km of the reactor, in the form of small particles of radioactive fuel ("hot particles").[40]

Initial concerns over safety of the food chain were primarily due to short-lived ^{131}I in milk and fresh vegetables.[41,42] Following the first few weeks after the accident, physical decay reduced activity concentrations of ^{131}I (and many other short half-life isotopes) to insignificant levels, and activity concentrations of deposited nuclides had declined significantly (see Figure 3). In the medium term after the accident, ^{137}Cs made up the major component of radiation doses to humans. Significant contributions to dose were made by the shorter half-life ^{134}Cs and also by ^{90}Sr.

Contamination of the food chain spread much wider than the evacuated areas, affecting many areas of Belarus, Ukraine and European Russia. Foodstuffs were contaminated in some areas of relatively low ^{137}Cs deposition as a result of high accumulation from certain soil types. For instance, Beresford and

Wright[43] report that in areas of soddy, podzolic soils, a ^{137}Cs deposition in the range of 140–500 kBq m^{-2} resulted in an annual ingested dose of 1 mSv. In areas of peaty soil, this level of dose was reached at ^{137}Cs depositions as low as 7–50 kBq m^{-2}. Parts of Western Europe were also affected, with advice not to consume fresh vegetables being given in, for example, parts of Italy and Germany. Most affected outside the former Soviet Union were the Scandinavian countries, where activity concentrations in reindeer, goat milk, sheep, game animals and freshwater fish were above intervention levels, and, as with sheep in some upland areas of the UK, were subject to long-term restrictions.[44]

A particular feature of the Chernobyl accident was the importance of wild foodstuffs in determining internal radiation doses. The collection of wild foodstuffs was and is important in the former Soviet Union, both as a pastime and a source of free food. In rural settlements surveyed within Belarus, Russia and the Ukraine, 40–75% of interviewees consumed wild fungi, 60–70% forest berries and 20–40% fish from local lakes.[45] As a result of the typically high uptake of radiocaesium to wild foodstuffs, it was estimated[46] that fungi and berries could contribute up to 60–70% of the dietary ^{137}Cs intake of those adults within Russia. Indeed, within the rural population in the affected parts of Russia a mean increase in the whole body radiocaesium activity of 60–70% in autumn as a result of fungi consumption was noted.[47]

The Chernobyl nuclear power plant is located on the Pripyat River which forms part of the Pripyat/Dnieper River and reservoir system providing drinking and irrigation water to approximately 15 million people in Ukraine. Though, after the initial few weeks following the accident, contamination of surface water systems was generally below drinking water limits,[48] major remediation works had to be put in place to demonstrate to a concerned population that water supplies were being protected. Bioaccumulation of radiocaesium in freshwater fish meant that guideline maximum levels (of approximately 1000 Bq kg^{-1}) were exceeded for many years in areas close to Chernobyl and in some parts of Western Europe.[48]

5.3 *Radiation Exposures and their Environmental and Health Impacts*

Initially, the radiation dose to humans was primarily from ^{131}I (with contributions from other short-lived isotopes) in the first few weeks after the accident. Over the months to decades after the accident, longer lived isotopes, primarily ^{137}Cs with contributions from ^{134}Cs and ^{90}Sr, formed the major part of the dose. Over hundreds to thousands of years, plutonium isotopes (with ingrowth of ^{241}Am) will form the major part of the dose as the ^{137}Cs and ^{90}Sr decay away (see Figure 3).

Measures to protect both the people on the site and the population of the surrounding areas were, in the very early stages of the accident, inadequate. Firemen had not been trained in radiation protection and had no dosimeters to control their radiation exposure. Although potassium iodide tablets (to block radioiodine uptake by the thyroid) were distributed to power plant workers within half an hour of the accident,[39] there was "no systematic distribution"[49] of tablets to the population of Pripyat, a town approximately 3 km from the plant. Face masks to help protect from inhalation of airborne radioactive particles were not available to the population and there were no official warnings for people to stay indoors, out of the contaminated air. Many children in Pripyat were playing outdoors during the day of the 26th of April (the accident occurred in the early hours of that day), unaware of the potential danger. Finally, there was no systematic ban on the consumption of milk, which led to high intakes of ^{131}I to the population during the weeks after the accident.

Once the full scale of the accident had been acknowledged, however, the local population was rapidly evacuated. At 14:00 on Sunday the 27th of April, the 44 000 population of Pripyat were evacuated in 1200 buses. On the 2nd of May, it was decided to evacuate people and cattle from an area of approximately 30 km radius around the plant (the "30 km Zone"), the boundary being based on a map of radiation dose rate. By the 6th of May, the entire 30 km Zone had been evacuated. Subsequent mapping of contamination later led to more evacuations, including areas in Belarus and the Bryansk region of Russia around 150 km to the north-west of the reactor. In total, approximately 116 000 people[50] and 60 000 cattle[39] were initially evacuated from an area of approximately 3500 km^2. In subsequent years many more people were evacuated from other contaminated areas, reaching a total of approximately 350 000. At present, many of the evacuated areas remain uninhabited, though some small areas have been re-settled.

In the early phase after the accident, 28 people died of radiation sickness. Amongst the 134 confirmed radiation sickness cases, a further 11 people died in the period 1987–98 from various causes including myelodysplastic syndrome (disorder of the bone marrow which has been linked to radiation), heart disease and cirrhosis of the liver.[39] Approximately 4000 cases of thyroid cancer were observed in children and adolescents during the period to 2003. These were overwhelmingly due to exposures to ^{131}I during the first weeks after the accident. Studies[51,52] have shown an increasing thyroid cancer risk with increasing radiation dose from short-lived ^{131}I. It is expected that thyroid cancer incidence will continue to be elevated in these groups as a result of their past exposure to ^{131}I. In a study in Belarus, Jacob *et al.*[52] observed 569 excess cases for the period 1991 to 1996 and predicted 12 000 excess cases for the period 1997–2036. These workers, however, noted that there was a large uncertainty in this prediction.

Increases in other solid cancers and leukaemia are also expected in the affected populations, though these are expected to be difficult to detect epidemiologically. The majority of excess leukaemia cases observed in survivors of the Hiroshima and Nagasaki atomic bombs occurred within 15 years of exposure.[53] No increased incidence of leukaemia has so far been observed in Ukrainian and Belarussian emergency workers. One study of Russian emergency workers,[41,54] however, observed an increase in the incidence of non-CLL leukaemia (non-Chronic Lymphocytic Leukaemia) between 1986 and 1996. Of the 71 217 Russian emergency workers studied, 21 had contracted non-CLL leukaemia. Approximately 50% of these cases were expected to have been radiation induced.[41]

Studies have been carried out of childhood leukaemia in the affected populations, though "the current information is scant and conclusions cannot be drawn about possible increases in childhood leukaemia following the Chernobyl accident".[55]

Based on known health impacts of radiation, increases in solid cancers (in addition to thyroid cancer) are expected following Chernobyl. However, due to the relatively low individual doses (and hence risks), these are expected to be difficult to detect epidemiologically. A study by the World Health Organisation[55] used the International Commission on Radiological Protection (ICRP) fatal cancer risk factor to estimate (as an "indication" of mortality impacts of Chernobyl), 9000 fatal cancers in the most exposed populations of Russia, Belarus and Ukraine. The collective dose to the World population resulting from the Chernobyl accident has been estimated to be 600 000 person-Sv,[56] resulting in an estimated potential fatal cancer total of 30 000.

It is clear that the extremely high levels of radioactivity deposited during and shortly after the accident damaged the ecosystem in parts of the 30 km Exclusion Zone. In an area of approximately 4 km^2 (the "Red Forest"), pine trees were killed shortly after the accident and serious damage to trees was observed over a much larger area (several tens of km^2). Though the direct evidence of fatalities in animals after the accident is sparse, it is likely that in the most highly contaminated areas fatalities occurred both through the direct effects of radiation and through damage to habitats. Reductions in rodent populations were observed in the most contaminated areas of the 30-km Zone during 1986, and increases in embryonic mortality were observed in 1986 and 1987. During 1987, however, rodent populations increased due in part to inward migration from less contaminated areas.

In the first few years after the accident, effects of radiation were observed on animals, for example on the liver of small mammals and some individuals had enlarged spleens. Dose rates to small mammals in the most contaminated areas were above those expected to cause reproductive damage. Some years after the accident, however, histological examinations of small rodents from the contaminated sites did not show significant levels of abnormalities.[57,58]

In assessing the ecological consequences of the Chernobyl accident, the negative impact of radiation on the environment must be weighed against the positive impact the removal of humans from the area has had on wildlife habitats. Twenty-five years after the accident there is some (often contradictory) evidence of continuing radiation damage to organisms, but this appears to be relatively minor (although also poorly understood). On a macro-ecological (*i.e.* large-) scale, however, there have been reports of a dramatic increase in populations of wild mammals and bird species living in the abandoned lands.[59,60]

The reports of a dramatic wildlife recovery at Chernobyl have been questioned by recent work showing apparently significant population level effects of radiation even (in terms of deterministic effects on animals) at very low dose rates.[61,62] Some of this work has been criticised by this author (J. T. Smith) for poor dosimetry, misleading descriptions of sample sites, and a failure to account for key differences between contaminated and control sites.[63] However, it is clear that further research is required at Chernobyl to test hypotheses of population-level impacts of chronic, low dose radiation in the environment.

5.4 Social and Psychological Impacts

The World Health Organization Chernobyl Forum Report[64] concluded that "The mental health impact of Chernobyl is the largest public health problem unleashed by the accident to date". Victims of radiation exposure have to live with the (to them, often unknown) future health risks of that exposure for the rest of their lives. Whilst to the scientific community these risks may be deemed low or insignificant compared to other risks encountered in daily life,[65] the exposure (and consequent evacuation) may do enormous psychological (and social) damage which in turn can have real health effects.

Populations in the contaminated areas have been shown to have higher levels of stress, worse perceived health and greater use of medical facilities (*e.g.* number of doctor visits) than similar unaffected populations. There have been media reports of large numbers of abortions carried out because of mothers' (unfounded) fear of radiation damage to their unborn child.

The social and psychological consequences of Chernobyl were exacerbated by the tendency of the Soviet government to conceal or downplay the seriousness of the accident. This reaction was, to a lesser extent, also seen in many western European countries. Once lost, public confidence in authorities is difficult to regain, and there is widespread mistrust in Ukraine, Belarus and Russia (and in Western Europe) of the pronouncements of government and scientists on radiation issues.

6 Conclusions

Insofar as the four accidents can be categorised, there is a clear dividing line between the circumstances leading to the early accidents at military facilities (Kyshtym and Windscale) and the later civilian nuclear power plant accidents (TMI and Chernobyl). In his review of Windscale and Kyshtym, Jones[2] identified the common factors in these accidents:

> *"...both occurred at installations whose main purpose was to produce plutonium for their respective national weapons programmes, at a time when pressures to produce the necessary material quickly were extreme; moreover in both cases the processes involved were imperfectly understood and would not be considered safe by modern standards".*

All four were failures both of equipment and management/operation of that equipment, but the latter two could both have been prevented had the operators taken the appropriate actions in the build up to and during the accidents. In fact, at both TMI and Chernobyl, it appears that the operators' actions, largely through no fault of their own, contributed significantly to the accident. An important contributing factor to the design and management failures leading to the TMI accident, identified by the President's Commission,[27] was an attitude that nuclear power plants were inherently safe; an attitude which also prevailed in the Soviet Union prior to Chernobyl. The President's Commission on TMI concluded that:

> *"The Commission is convinced that this attitude must be changed to one that says nuclear power is by its very nature potentially dangerous, and, therefore, one must continually question whether the safeguards already in place are sufficient to prevent major accidents".*

> *"We are convinced that if the only problems were equipment problems, this Presidential Commission would never have been created. The equipment was sufficiently good that, except for human failures, the major accident at Three Mile Island would have been a minor incident. But, wherever we looked, we found problems with the human beings who operate the plant, with the management that runs the key organization, and with the agency that is charged with assuring the safety of nuclear power plants".*[27]

Since both TMI and Chernobyl, major improvements have been made in power plant design and in the safety culture of the nuclear industry. But the recent accident at the Fukushima Daiichi nuclear power plant serves as a reminder that extreme events can and do happen. The nuclear industry and regulators must not allow the belief to take hold that major accidents are impossible. It was a major earthquake and Tsunami which caused the Fukushima accident, but a human planning failure player a important part.

In terms of environmental and human health impacts, it is obvious that releases of radioactive materials at Kyshtym and Chernobyl had major impacts on the human population both in terms of enhanced cancer risk and, importantly, the social, psychological and economic impacts of permanent evacuation. Though many radiation-induced cancers, even from the Chernobyl accident, are never likely to be epidemiologically distinguishable from "natural" background cancers, it is possible to estimate the cancer effects from estimates of collective dose. Indicative estimates of collective doses from the four accidents is given in Table 4, though this comparison is far from comprehensive, partly because of the difficulty in estimating doses, and partly because different approaches were used in the different studies. Despite these limitations, it is clear the collective doses from Chernobyl were by far the most significant. It is important to note, however, that the influence of atmospheric nuclear weapons testing on the collective dose to the World population was much greater than that for Chernobyl. All of these collective doses are dwarfed by the much greater collective doses to the World population from natural and medical sources of radiation.

The damage to the ecosystem caused by these accidents was severe in small areas where organisms were exposed to extremely high doses in the period following the releases. However, long-term environmental damage from chronic,

Table 4 Summary of estimated collective effective dose from each accident in comparison with that from atmospheric nuclear weapons testing.

Accident	*Estimated collective effective dose (person-Sv)*
Windscale	2000
Kyshtym	>9500[a]
Three-Mile Island	15
Chernobyl	600 000
Atmospheric Nuclear Weapons Testing	30 000 000[b]

[a]This includes only the exposures to the evacuated population and clean up workers during their period in the contaminated area and does not include doses to people outside the evacuated area, or to the 5000 workers on-site at the time of the accident.
[b]From UNSCEAR.[73]

lower level, radiation is less clear. Evidence of long-term damage to organisms (at a genetic, individual or population level) from these studies is often contradictory, partly as a result of poor study design and methods in some studies, but also because of the confounding from other environmental and ecological variables. At both Chernobyl and Kyshtym, the evacuated areas have, in the long term after the accident, been described by some as a "nature reserve" since the damage human influence has on ecosystems has been removed.

References

1. J. A. Garland and R. Wakeford, *Atmos. Environ.*, 2007, **41**, 3904–3920.
2. S. Jones, *J. Environ. Radioact.*, 2008, **99**, 1–6.
3. A. C. Chamberlain, *Q. J. R. Meteorol. Soc.*, 1959, **85**, 350–361.
4. T. Bergan, M. Dowdall and Ø. G. Selnæs, *J. Environ. Radioact.*, 2008, **99**, 50–61.
5. A. H. Wolff, *Publ. Health Rep.*, 1959, **74**, 42–43.
6. J. A. Auxier, in *AMA Toxic Agents Conference*, Washington DC, USA, 1987, pp. 80–84.
7. D. Jackson and S. Jones, in *Comparative Assessment of the Environmental Impact of Radionuclides Released During Three Major Nuclear Accidents: Kyshtym, Windscale and Chernobyl*, European Commission, Luxembourg, 1991, pp. 1015–1040.
8. H. J. Dunster, H. Howells and W. L. Templeton, *2nd International Conference on Peaceful Uses of Atomic Energy*, New York, 1958.
9. S. Wright, J. Smith, N. Beresford and W. Scott, *Radiat. Environ. Biophys.*, 2003, **42**, 41–47.
10. M. J. Crick and G. S. Linsley, *Int. J. Radiat. Biol.*, 1984, **46**, 479–506.
11. R. H. Clarke, in *Medical Reponse to Effects of Ionising Radiation*, ed. W. A. Crosbie and J H. Gittus, Elsevier Applied Science, Amsterdam, 1989, pp. 102–118.

12. D. McGeoghegan and K. Binks, *J. Radiolog. Protect.*, 2000, **20**, 261.
13. B. Wynne, *Publ. Understanding Sci.*, 1992, **1**, 281–304.
14. F. Fremont-Smith, *Am. J. Orthopsychiat.*, 1958, **28**, 456–466.
15. D. L. Collins, *Nuclear Accidents in the Former Soviet Union: Kyshtym, Chelyabinsk and Chernobyl,* Armed Forces Radiobiology Research Institute, Bethesda, MD, USA, 1991.
16. B. V. Nikipelov, G. N. Romanov, L. A. Buldakov, N. S. Babaev, Y. B. Kholina and E. I. Mikerin, *At. Energy*, 1989, **67**, 569–576.
17. R. M. Alexakhin, S. V. Fesenko and N. I. Sanzharov, *Radiat. Protect. Dosimetry*, 1996, **64**, 37–42.
18. B. V. Nikipelov, G. N. Romanov, L. A. Buldakov, N. S. Babaev, Y. B. Kholina and E. I. Mikerin, *Accident in the Southern Urals on 29 September 1957,* International Atomic Energy Agency, Vienna, Austria, 1989.
19. M. A. Cross, University of Portsmouth, PhD Thesis, 2001.
20. K. Muck, M. Sinojmeri, H. Whilidal and F. Steger, *Radiat. Protect. Dosimetry*, 2001, **94**, 251–259.
21. I. L. Abalkina, A. A. Sarkisov, I. I. Linge, S. V. Kazakov, S. V. Panchenko and E. A. Savelieva, *Appl. Radiat. Isot.*, 2008, **66**, 1554–1557.
22. M. M. Kossenko, *Radiat. Protect. Dosimetry*, 1995, **62**, 87–89.
23. V. Pozolotina, I. Molchanova, E. Karavaeva, A. Aarkrog and S. P. Nielsen, *Radiation Exposures by Nuclear Facilities. Evidence of the Impact on Health*, eds. I. Schmitz-Feuerhake and M. Schmidt, Gesellschaft für Strachlenschutz, Berlin, 1996, pp. 382–386.
24. D. A. Krivolutsky, in *Proceedings of the NATO Advanced Research Workshop on Bioindicator Systems for Soil Pollution*, ed. N. M. van Straalen and D. A. Krivolutsky, Kluwer Academic Publishers, Moscow, Russia, 1995, pp. 189–196.
25. I. I. Kryshev and T. G. Sazykina, *Radiat. Protect. Dosimetry*, 1998, **75**, 187–191.
26. NRC, *US. Nuclear Regulatory Commission*, 2010, www.nrc.gov.
27. J. G. Kemeny, *Report of the President's Commission on the Accident at Three Mile Island*, Pergamon, New York, NY, USA, 1979.
28. C. W. Miller, S. J. Cotter, R. E. Moore and C. A. Little, *Estimates of Dose to the Population within Fifty Miles due to Noble Gas Releases from the Three Mile Island Incident*, Oak Ridge National Laboratory, Oak Ridge, TN, USA, 1980.
29. M. C. Hatch, J. A. N. Beyea, J. W. Nieves and M. Susser, *Am. J. Epidemiol.*, 1990, **132**, 397–412.
30. T. M. Gerusky, *Ann. N.Y. Acad. Sci.*, 1981, **365**, 54–62.
31. R. L. Katheren, *Radioactivity in the Environment: Sources, Distribution, and Surveillance,* Harwood Academic Publishers, New York, NY, USA, 1984.
32. R. W. Field, *Radiat. Protect. Dosimetry*, 2005, **113**, 214–217.
33. S. Wing, D. Richardson, D. Armstrong and D. Crawford-Brown, *Environ. Health Perspect*, 1997, **105**, 52–57.
34. E. O. Talbott, A. O. Youk, K. P. McHugh-Pemu and J. V. Zborowski, *Environ. Health Perspect*, 2003, **111**, 341–348.

35. B. A. Osif, A. J. Baratta and T. W. Conkling, *Three Mile Island 25 Years Later*, Penn State University Press, PA, USA, 2004.
36. C. Susan and B. Kent, *Disasters*, 1982, **6**, 116–124.
37. R. Fleming, A. Baum, M. M. Gisriel and R. J. Gatchel, *Journal Name: J. Human Stress*, 1982, **8**(3), 14–22.
38. A. Baum, R. J. Gatchel and M. A. Schaeffer, *J. Consulting Clin. Psychol.*, 1983, **51**, 565–572.
39. UNSCEAR, *Report to the General Assembly: Sources and Effects of Ionizing Radiation. Volume II, Annex J.*, United Nations, New York, NY, USA, 2000.
40. K. Muck, G. Prohl, I. Likhtarev, L. Kovgan, R. Meckbach and V. Golikov, *Health Phys.*, 2002, **82**, 141–156.
41. A. F. Tsyb, V. K. Ivanov, V. A. Sokolov, A. I. Gorski, M. A. Maksioutov, O. K. K. Vlasov, S. E. and A. M. Godko, *Radiat. Risk*, 2002, Special Issue: Health consequences 15 years after the Chernobyl catastrophe: data of the National Registry, 39–50.
42. V. V. Drozdovitch, G. M. Goulko, V. F. Minenko, H. G. Paretzke, G. Voigt and J. I. Kenigsberg, *Radiat. Environ. Biophys.*, 1997, **36**, 17–23.
43. N. A. Beresford and S. M. Wright, *Self-Help Countermeasure Strategies for Populations Living within Contaminated Areas of the Former Soviet Union and an Assessment of Land Currently Removed from Agricultural Usage*, Institute of Terrestrial Ecology, Grange-over-Sands, UK, 1999.
44. B. J. Howard, N. A. Beresford and G. Voigt, *J. Environ. Radioact.*, 2001, **56**, 115–137.
45. P. Strand, B. J. Howard and V. Averin, *Transfer of Radionuclides to Animals, their Comparative Importance under Different Agricultural Ecosystems and Appropriate Countermeasures. Experimental Collaboration Project No. 9 Final Report*, European Commission, Luxembourg, 1996.
46. V. N. Shutov, G. Y. Bruk, L. N. Basalaeva, V. A. Vasilevitskiy, N. P. Ivanova and I. S. Kaplan, *Radiat. Protect. Dosimetry*, 1996, **67**, 55–64.
47. L. Skuterud, I. G. Travnikova, M. I. Balonov, P. Strand and B. J. Howard, *Sci. Total Environ.*, 1997, **193**, 237–242.
48. J. T. Smith and N. A. Beresford, *Chernobyl – Catastrophe and Consequences,* Springer-Praxis, Berlin, Germany, 2005.
49. IAEA, *The International Chernobyl Project Technical Report*, International Atomic Energy Agency, Vienna, Austria, 1991.
50. S. T. Belyaev, V. F. Demin, V. A. Kutkov, V. G. Bariakhtar and E. P. Petriaev, *The Radiological Consequences of the Chernobyl Accident*, 1996.
51. P. Jacob, G. Goulko, W. F. Heidenreich, I. Likhtarev, I. Kairo, N. D. Tronko, T. I. Bogdanova, J. Kenigsberg, E. Buglova, V. Drozdovitch, A. Golovneva, E. P. Demidchik, M. Balonov, I. Zvonova and V. Beral, *Nature*, 1998, **392**, 31–32.
52. P. Jacob, Y. Kenigsberg, G. Goulko, E. Buglova, F. Gering, A. Golovneva, J. Kruk and E. P. Demidchik, *Radiat. Environ. Biophys.*, 2000, **39**, 25–31.
53. D. A. Pierce, Y. Shimizu, D. L. Preston, M. Vaeth and K. Mabuchi, *Radiat. Res.*, 1996, **146**, 1–27.

54. A. P. Konogorov, V. K. Ivanov, S. Y. Chekin and S. E. Khait, *J. Environ. Pathol., Toxicol. Oncol.*, 2000, **19**, 143–151.
55. E. Cardis, G. Howe, E. Ron, V. Bebeshko, T. Bogdanova, A. Bouville, Z. Carr, V. Chumak, S. Davis, Y. Demidchik, V. Drozdovitch, N. Gentner, N. Gudzenko, M. Hatch, V. Ivanov, P. Jacob, E. Kapitonova, Y. Kenigsberg, A. Kesminiene, K. J. Kopecky, V. Kryuchkov, A. Loos, A. Pinchera, C. Reiners, M. Repacholi, Y. Shibata, R. E. Shore, G. Thomas, M. Tirmarche, S. Yamashita and I. Zvonova, *J. Radiol. Protect.*, 2006, **26**, 127–140.
56. B. Bennett, in *One Decade after Chernobyl: Summing up the Consequences of the Accident*, IAEA, Vienna, Austria, 1996, pp. 117–126.
57. R. J. Baker, J. A. Dewoody, A. J. Wright and R. K. Chesser, *Ecotoxicology*, 1999, **8**, 301–309.
58. D. Jackson, D. Copplestone and D. M. Stone, *Nucl. Energy*, 2004, **43**, 281–287.
59. N. Williams, *Science*, 1995, **269**, 304.
60. R. J. Baker and R. K. Chesser, *Environ. Toxicol. Chem.*, 2000, **19**, 1231–1232.
61. A. P. Møller, T. A. Mousseau, G. Milinevsky, A. Peklo, E. Pysanets and T. SzÉP, *J. Animal Ecol.*, 2005, **74**, 1102–1111.
62. A. P. Møller, T. A. Mousseau, F. de Lope and N. Saino, *Biol. Lett.*, 2007, **3**, 414–417.
63. J. T. Smith, *Biol. Lett.*, 2008, **4**, 63–64.
64. WHO, *Health Effects of the Chernobyl Accident and Special Health Care Programmes,* World Health Organization, Geneva, Switzerland, 2006.
65. J. Smith, *BMC Public Health*, 2007, **7**, 49.
66. C. A. Johnson, K. P. Kitchen and N. Nelson, *Atmos. Environ.*, 2007, **41**, 3921–3937.
67. G. J. Dicus, in *Joint Meeting of American Nuclear Society, Washington, D.C. Section and Health Physics Society, Baltimore-Washington Chapter*, U.S. Nuclear Regulatory Commission, Washington D.C., DC, USA, 1997.
68. M. De Cort, *Atlas of Caesium Deposition on Europe after the Chernobyl Accident. EUR 16733,* European Commission, Luxembourg, 1998.
69. P. H. Gudiksen, T. F. Harvey and R. Lange, *Health Phys.*, 1989, **57**, 697–706.
70. R. S. Cambray, K. Playford, G. N. J. Lewis and R. C. Carpenter, *Radioactive Fallout in Air & Rain, Results to the End of 1988 AERE R 13575,* Atomic Energy Authority, London, UK, 1989.
71. NCI, *Estimated Exposures and Thyroid Doses Received by the American People from Iodine-131 in Fallout Following Nevada Atmospheric Nuclear Bomb Tests*, U.S. National Cancer Institute, Bethesda, MD, USA, 1997.
72. J. Gray, S. R. Jones and A. D. Smith, *J. Radiol. Protect.*, 1995, **15**, 99.
73. UNSCEAR, *Sources and Effects of Ionizing Radiation*, United Nations Scientific Committee on the Effects of Atomic Radiation UNSCEAR 1993 Report to the General Assembly, with Scientific Annexes, United Nations, New York, USA, 1993.
74. A. Kruuglov, *The History of the Soviet Atomic Industry*, Taylor & Francis, London, 2002.

Management of Land Contaminated by the Nuclear Legacy

RICHARD KIMBER,* FRANCIS R. LIVENS AND
JONATHAN R. LLOYD

ABSTRACT

The widespread spread use of nuclear materials over the past 60 years has lead to anthropogenic release of radionuclides into the environment. The release of such contaminants is currently of great public concern and scientific interest worldwide. Contamination has arisen on sites involved in both military and civilian uses of nuclear material through leakages, spills, controlled discharges and munitions use. The management of this nuclear legacy is a global priority as governments seek to decommission and reclaim land contaminated by the use of nuclear facilities. The scale of contamination presents a serious financial burden with the cleanup of US sites expected to cost up to a trillion dollars. In the UK, the problem exists on a smaller but significant scale with associated cleanup costs estimated to be in the order of £100 billion. A wide range of disciplines are required to understand the behaviour of radionuclides and co-contaminants in these contaminated environments in order for effective remediation techniques to be utilised. Potential remediation strategies cover a range of biological, chemical and physical methods which can be used to treat the complex contamination scenarios found at nuclear sites. A number of these remediation techniques have been trialled at several sites managed by the United States Department of Energy with some success in treating radionuclide contamination.

* Corresponding author

Issues in Environmental Science and Technology, 32
Nuclear Power and the Environment
Edited by R.E. Hester and R.M. Harrison

Published by the Royal Society of Chemistry, www.rsc.org

1 Introduction

The industrial production of nuclear materials and their subsequent use, both militarily and civilian, since the 1940s has left behind a legacy of contamination and hazardous waste. Management of this legacy is now a global priority as many governments seek to decommission and reclaim land once used by their nuclear facilities. However, the storage and disposal of contaminated equipment and spent nuclear material, compounded with the widespread dispersion of radionuclides and co-contaminants at these sites makes this a challenging and expensive undertaking.

Many former nuclear facilities were shut down at the end of the Cold War in the 1990s as the demand for nuclear weapon production decreased. These sites are now the focus of remediation, decommissioning and decontamination efforts for governments and agencies worldwide. Substantial quantities of land, groundwater, and equipment have been contaminated by the former operations of these sites, the details of which will be discussed in this chapter. A number of sites which are still in operation, often dealing with the reprocessing of nuclear material, are also the focus of ongoing remediation efforts. Contamination issues have arisen from various sources including accidental release, the controlled discharge of nuclear waste and the use of radionuclide-containing munitions. The accidental release of radionuclides can occur through the leakage and spills of radioactive material, as well as from incidents (such as explosions) which have occurred on site. Off site contamination is a concern in cases where both on site accidents and natural transport processes have spread radionuclides further afield. Hydraulic flow presents an ongoing problem for the containment of on site contaminants as groundwater plumes threaten to spread contamination to aquifers used in the irrigation of crops or for public drinking water. For this reason, the mobility of radionuclides, heavy metals and toxic organics is a key factor in determining the risk that each contaminant presents to the environment and general public. Understanding the mechanisms which affect contaminant mobility is therefore vital in developing effective remediation strategies. Numerous techniques are available for treating contaminated land and groundwater, generally falling under either biological (bioremediation) or chemical processes. This chapter will outline some of the key techniques available, along with their associated advantages and disadvantages. A number of key case studies relating to former nuclear facilities will also be discussed where a variety of techniques have been applied in field scale studies.

2 Contamination at Worldwide Nuclear Facilities

In 2008, 439 nuclear power plants were in operation in 31 countries accounting for 15% of the world's electricity production[1] and as of January 1999, a total of 2532 nuclear detonations have been performed.[2] Many sites involved in the production, reprocessing and storage of nuclear materials have contributed to environmental radionuclide release through discharge, spillage, accidents and testing facilities. A summary of such sites and their associated contamination issues is given in Table 1. The number of nuclear facilities is expected to increase in the near future as governments renew their interest in nuclear power due its reduced CO_2 emissions and the energy security it provides when compared to other forms of electricity generation.

2.1 United Kingdom

The United Kingdom's nuclear legacy has arisen from a variety of nuclear facilities operated across the country over the past ~60 years. These contribute to the production of nuclear material for nuclear reactors or weapons, the use of this material in reactor plants and the re-processing of spent nuclear fuel. Construction on the UK's first nuclear power plant, Calder Hall, began in 1953 and in 1956 it was connected to the national grid becoming the world's first commercial nuclear power station. The site was also expanded over the following decades, to result in the present Sellafield site (see section 2.1.1). Between 1953 and 1971 a total of 26 reactors were built at nuclear research and development sites across the UK.[3] A substantial part of the UK's electricity supply has come from the first generation of Magnox nuclear power stations over the past 60 years. Just two of the eleven Magnox stations are still operational.

The Atomic Weapons Establishment (AWE) is responsible for providing and maintaining the UK's nuclear deterrent and has held this responsibility for over 50 years. AWE operates over two sites, Aldermaston which is a former airfield and Burghfield, a former munitions factory. Although the Aldermaston site is radiologically safe there are areas where soils contain higher than background levels of various radionuclides, including plutonium. Levels of $^{239+240}Pu$ have been found to range from 15 to 155 Bq kg^{-1} in certain settled sediments (sludge)[4] compared to background levels due to global fallout of 0.02 to 0.7 Bq kg^{-1}.[5]

The Nuclear Decommissioning Authority (NDA), a Non-Departmental Public Body (NDPB), was established in 2005 to manage the decommissioning and clean up of the UK's civil public sector nuclear legacy sites. The restoration program tasked to the NDA relates to 19 sites covering the length and breadth of the UK with certain sites not expected to reach their planned end state for decades. The discounted lifetime cost for completing their contracted work, the Nuclear Liabilities Estimate (NLE), stands at £44.5 billion.[3] A detailed overview of the NDA's planned approach for decommissioning and clean up is provided in their recent draft strategy published for consultation.[3]

Although there are a number of sites in the UK where nuclear operations have occurred, the majority of the legacy waste and contamination is located at a few principal facilities. Two key sites (Sellafield and Dounreay) involved in the UK's nuclear waste inventory and which suffer from the greatest contamination concerns are discussed here in more detail.

2.1.1 Sellafield

Sellafield (formerly Windscale), West Cumbria, is the UK's largest nuclear complex covering 262 hectares and has supported the nuclear power program

Table 1 Summary of former and current uses of key worldwide nuclear facilities with associated soil and groundwater contamination issues. (Adapted from W. Standring, M. Dowdall and P. Strand, *Int. J. Environ. Public Health*, 2009, **6**, 1).

Site	*Former uses*	*Current use*	*Soil contamination*	*Groundwater contamination*	*Sources*
Sellafield	WW2 explosives production site. Postwar plutonium production, fuel reprocessing, fuel manufacturing, nuclear waste management, nuclear energy generation	Reprocessing, fuel production, waste management and decommissioning	~1600 m^3 contaminated with radioactive waste (ILW). ~1 000 000 m^3 will require treatment as LLW.	Monitoring results from 2009 • Total alpha activity below WHO safe drinking water limit (0.5 Bq l^{-1}) in all but 5 cases • Total beta activity above WHO safe drinking water limit (1 Bq l^{-1}) with a highest annual average being 129 000 Bq l^{-1}. Groundwater plumes contain ^{90}Sr, ^{137}Cs, ^{3}H and ^{99}Tc.	ref. 8, 10, 11
AWE Aldermaston/ Burghfield	Former RAF site/former munitions site Production/service/ decommissioning of nuclear weapons	Maintenance and decommissioning of nuclear weapons	$^{239+240}Pu$ on settled sediment varies from 15–155 Bq kg^{-1}.	Run off water contains: • 0.7 to 44 μBq kg^{-1} dissolved $^{239+240}Pu$ • 1.2 to 400 Bq kg^{-1} particle bound $^{239+240}Pu$.	ref. 4
Dounreay	RAF site converted to UK's centre for fast reactor R&D	All reactors closed down, site is in final decommissioning stage (closure expected 2032)	Irradiated fuel particles found on local beaches. ^{137}Cs contamination on site in small pockets of >4 Bq g^{-1}.	Legacy of irradiated particles discharged into sea during 1960/70s. Particles detected on seabed around Dounreay.	ref. 14, 15

Table 1 Continued.

Site	*Former uses*	*Current use*	*Soil contamination*	*Groundwater contamination*	*Sources*
Mayak	Nuclear weapons production, reprocessing nuclear material	Produces radioisotopes and electrical equipment for monitoring, reprocesses fuel, decommissioning	HLW tank exploded in 1957 • 740 PBq released • –90% contaminated immediate vicinity. • Parts of the Mayak site have dose rates up to 15 mR h^{-1}.	Open reservoirs on site contain 340 million m^3 of radioactive water. In 1993: • 70 MBq l^{-1} ^{90}Sr and 100 MBq l^{-1} ^{137}Cs in Reservoir 9, (Lake Karachay, R9) were measured • groundwater pollution plume from R9 covers 10 km^2, spreading at 80–100 m y^{-1}.	ref. 20 Standring, 2009
Rocky Flats	Manufacture of nuclear weapons	Cleanup and closure completed in 2005. Most of the land transferred to US Fish and Wildlife Service 2007 for use as reservation	Soils from a contaminated toposequence contain activities of: • $^{239+240}Pu$ in the range 2220 to 11 460 Bq kg^{-1} • ^{241}Am in the range 1840 to 8840 Bq kg^{-1}.		ref. 31

Oak Ridge	Separation of uranium for Manhattan Project	National Lab managed by DoE	Sorbed and precipitated uranium concentrations up to 800 mg kg^{-1}. Hg up to 2400 μg g^{-1} in floodplains along East Fork Popular Creek.	Soluble uranium in groundwater plume (up to 210 μM). Leakage from S-3 ponds has created a plume containing uranium (up to 0.2 μM) and Tc (up to 47 nM).	ref. 38, 39
Hanford	Plutonium production, nuclear reactors	Decommissioning and cleanup	68 out of 149 tanks known or thought to have leaked HLW into sediments beneath them. Pu found in silt layers at up to 9.25 MBq kg^{-1}. Caesium-137 as high as 10^5 Bq g^{-1} in contaminated sediments.	In 1951, 3.5×10^5 l of highly radioactive waste leaked into subsurface containing an estimated 7000 kg of U. Tritium and ^{129}I present in groundwater at above drinking level limits. Tc, U, Pu ^{60}Co, ^{137}Cs also detected above drinking levels.	ref. 42, 44, 49
Rifle	Former uranium processing	UMTRA managed site		Uranium concentrations in a contaminated aquifer range from 0.4 to 1.4 μM	ref. 28

since the 1940s with the site containing the world's first commercial nuclear power station, Calder Hall. Operations at the Sellafield site include spent fuel reprocessing, mixed oxide fuel fabrication (MOX) and nuclear waste storage and management. Discharges into the environment from Sellafield began in 1951 and first became subject to formal authorisation in August 1954 under the "Atomic Energy Authority Act 1954". Prior to 1954, discharges were subjected to controls derived from consultation with site operators and government departments. Current disposal of radioactive waste is regulated under the "Environmental Permitting (England and Wales) Regulations 2010" (EPR).

During reprocessing, plutonium, uranium and highly radioactive fission products are separated by a series of solvent extractions which results in some of these products being concentrated in aqueous waste. Highly radioactive aqueous waste is added to an acid effluent stream for evaporation and storage and is now being converted into vitrified waste. Low level aqueous waste is discharged into the Irish Sea *via* pipelines extending 2.5 km from the high water mark. These low level discharges have created an environmental inventory, over the period of 1952–1990, of around 1.1×10^2 TBq of ^{238}Pu, 6.1×10^2 TBq of ^{239,240}Pu, 1.3×10^4 TBq of ^{241}Pu and 9.4×10^2 TBq of ^{241}Am (with about 3.6×10^2 TBq of the americium having been derived from decay of ^{241}Pu released).[6] Around 90% of the Pu, in its insoluble Pu(IV) state, was retained rapidly by the sediment in the Irish Sea along with the vast majority of the discharged Am. The remaining 10% of plutonium, in the more soluble Pu(V) state, remained in solution and was transported out of the Irish Sea.[7] Since 2006, beach monitoring has detected a number of contaminated sites resulting from the Sellafield discharges, although they are generally less active than those found in Dounreay.[8]

Approximately 1600 m^3 of soil around the centre of Sellafield has been contaminated by spillage and reprocessing and will have to be treated as intermediate level waste (ILW).[8] This area overlies an aquifer in the underlying sandstone geology, which is significantly contaminated to the southwest due to leaching of the contaminated soil from above. An estimated 1 000 000 m^3 of soil will require treatment as low level waste (LLW). Sellafield is also responsible for the storage of the majority of the UK's nuclear waste products and, as such, a large inventory of varying levels of radioactive waste is stored on site either awaiting disposal[9] or for the activity to decrease.

Two site investigations have been conducted at Sellafield over the past decade in an attempt to identify and develop conceptual models of below ground contamination. The first phase of the report was completed in 2004 and examined contamination outside of the Sellafield "Separation Area", where fuel re-processing and fabrication took place, with the second report focussing on contamination within the Separation Area expected to be completed in 2010.

Soil sample records from over 2000 boreholes have demonstrated that radioactively contaminated ground exists beneath and, occasionally, outside the Separation Area. Groundwater monitoring throughout the site has revealed that radioactive contamination is present in distinct plumes in the groundwater which are migrating in the direction of the hydraulic gradient. These contaminants include ^{90}Sr, ^{137}Cs, ^{3}H and ^{99}Tc, with actinides also expected.[10]

The maximum activity of the most mobile contaminant, tritium, is around 1.0×10^7 Bq m^{-3} in contaminated groundwater found in boreholes close to the Separation Area. The activity decreases down the hydraulic gradient towards the River Ehen, until it becomes undetectable (below 1.0×10^5 Bq m^{-3}).[10] Technetium-99, although derived from a different source, becomes a co-contaminant with the tritium in a common plume as they both migrate downgradient. The ^{99}Tc is known to be a contaminant in the upper strata of the sandstone bedrock and has also been found in monitoring wells as far as the site boundary. The maximum concentration of ^{99}Tc found in this plume during the phase 1 site investigation was 2.3×10^5 Bq m^{-3}, located near to the site main gate.[10] Strontium-90, which has limited solubility and readily adsorbs to sediments at Sellafield, is detectable in monitoring wells inside the Separation Area where it is mostly contained. Beta activity from the ^{90}Sr is also detected in two plumes, including the plume contaminated with ^{3}H and ^{99}Tc.[10] Caesium-137, the only other radioactive isotope detected in the groundwater plumes, was found to be present only in very low concentrations and only in filtered solids.

Monitoring of 137 boreholes was conducted for the Sellafield Ltd *Groundwater Annual Report*[11] and is summarised in Table 2. Although the majority of boreholes contain activity below the WHO drinking standard for total alpha, tritium and technetium activity, there are a significant number of boreholes with total beta activity above the WHO drinking standard. Strontium-90 makes up the bulk of the total beta activity, with caesium-137 also contributing significant activity. However, when both isotopes are examined on an individual

Table 2 Summary of the groundwater monitoring of 137 Sellafield boreholes.[11]

Activity analysed	*WHO drinking water standard (Bg l^{-1})*	*Boreholes where WHO standard is exceeded*	*Location of boreholes*	*Major isotopes*	*Highest annual average activity (Bq l^{-1})*
Total alpha activity	0.5	5	Within the Separation Area	Uranium isotopes	103
Total beta activity	1	46	Predominately within Separation Area with several to the south. A minority found close to the River Calder's west bank	^{90}Sr and ^{137}Cs	129 000
Tritium	10 000	3	Outside south-west corner of Separation Area	^{3}H	39 200
Technetium	100	1	Between south-west corner of Separation Area and the site main gate.	^{99}Tc	111

basis, then fewer samples exceed the WHO drinking standard for ^{90}Sr and no samples exceed the ^{137}Cs safe drinking limit. The majority of boreholes with values above the WHO standard are located within the Separation Area with a number also located to its south-west.

The former storage and de-canning facility, known as B-30, houses a pond used for the storage of spent nuclear fuel until its replacement facility, the Fuel Handling Plant, was commissioned in 1986. Although now closed, the storage pond is thought to contain 300 to 450 tonnes of spent nuclear fuel. Fuel was stored in the pond for longer than was anticipated due to an accident at the Magnox reprocessing facility in 1974 causing corrosion of the fuel cans and leakage of radiation into the pond.

2.1.2 Dounreay

Another former RAF site, Dounreay, became the centre for the UK's fast reactor research and development in 1954. Commercial energy production began in 1962 becoming the first fast reactor in the world to supply energy to the grid. However, fast reactor technology proved to be more expensive than was first thought and consequently all fast reactor programs ceased operations in 1994. Reprocessing and fuel fabrication operations ended in 1996 and 2004, respectively. Dounreay is now wholly a decommissioning site owned by the NDA and run by Dounreay Site Restoration Ltd. The site closure program is scheduled to be completed by 2025 at an estimated cost of £2.6 bn. Over the course of decommissioning, Dounreay is expected to generate a lifetime waste of 97 126 m^3 of LLW, 3164 m^3 of ILW and 0 m^3 of high level waste (HLW).[12] Dounreay has a legacy of irradiated nuclear fuel particles which were discharged into the sea as a result of reprocessing activities during the 1960s and 1970s. These particles have been detected on the seabed around Dounreay with the most hazardous fragments located close to the old discharge point on the seabed. Their disintegration is believed to be the source of smaller, less hazardous particles detected on local beaches. Around 1000 significant (10^6 Bq of ^{137}Cs), 1000 relevant (10^5 to 10^6 Bq of ^{137}Cs) and 3000 minor ($<10^5$ of ^{137}Cs) particles are thought to be present within the main particle plume offshore from Dounreay.[13] Monitoring of the particles is expected to last until 2020s and with a total cost estimated at £18–25 million.

On site, there are pockets of caesium-137 contamination with activities greater than 4 Bq g^{-1}, although the majority of contamination over the site is below 0.4 Bq g^{-1}.[14] Between 1959 and 1971, solid ILW was disposed of in the Dounreay waste shaft. A purpose built wet silo was constructed in 1971 as an alternative to the shaft after which solid ILW was tipped into the silo, a large underground concrete vault filled with water. Large items too big for a purpose built wet silo continued to be disposed of down the shaft until 1977, when an explosion in the airspace above the water column damaged the shaft cover. There are uncertainties over the exact contents of the shaft, thought to include contaminated equipment, chemicals, natural uranium fuel, radioactive sources and sludges.[15] A total of 703 m^3 of waste is covered by a water column 8 m deep which is below sea level so that groundwater flow is towards the shaft.

2.2 Russia

A large and complicated nuclear legacy has been left behind by the break up of the Soviet Union with numerous nuclear facilities located in Russia and other former Soviet states. Although civilian activities have contributed to this legacy, the majority of contamination issues in the former Soviet Union were created by military nuclear facilities used for the production of nuclear weapons. This problem was exacerbated by a previously relaxed attitude towards environmental issues with regards to nuclear waste disposal. Three nuclear facilities, Chelyabinsk-65, Tomsk-7, and Krasnoyarsk-26 operated in secret in the Ural mountains during the Cold War and were not subject to strict environmental practices.[16] Of these sites, Chelyabinsk-65 (Mayak) is the most publicised regarding its former activities, revealing a long history of accidental release and discharges, contributing to significant environmental contamination which will be discussed in greater detail below.

The worst nuclear power plant accident in history, the only level 7 event on the International Nuclear Event Scale to have occurred, happened on the 26th of April 1986 at Chernobyl, when a test was carried out to determine the ability of a turbine generator to provide power in the event of a station blackout. Serious violations of safety procedures and operating rules resulted in a steam explosion, cutting cooling channels on both sides of the reactor core resulting in a further explosion.[17] The release of ^{137}Cs from the explosion is estimated to have been around 85 PBq with an estimated 1760 PBq of ^{131}I, 10 PBq of ^{90}Sr and 3 PBq of plutonium isotopes also released.[18,19]

2.2.1 Mayak

Mayak, formerly known as Chelyabinsk-40 and later as Chelyabinsk-65, is one of the biggest nuclear facilities in the Russian Federation housing the former Soviet Union's first industrial nuclear reactors. The facility was responsible for producing the material for the country's first atomic bomb beginning in 1948.

Between the commencement of operations in 1948 through to September 1951, 78 million m^3 of high-level nuclear waste containing a total of 10^{17} Bq of beta activity was discharged from the radiochemical plant directly into the Techa River six kilometres below its source.[20] A radiation survey in 1951, revealed extensive contamination of the floodplain and bed of the Techa River and consequently excessive exposure to the inhabitants of the region. Of the total radioactivity discharged into the Techa, 99% was deposited in the first 35 km downstream. Four reservoirs were created along the Techa below Lake Kzyzyltash to isolate the most contaminated water. The final reservoir was completed in 1964 and including Lake Kyzyltash, now contains a volume of 380 million m^3 and about 7141 TBq of ^{90}Sr and ^{137}Cs.[20]

Discharge of diluted HLW directly into the Techa stopped in 1951 and instead was diverted into Lake Karachay. In 1953, an intermediate waste storage tank was put into operation but the excess supernatant (containing much of the caesium waste) was still discharged into the lake (29.6 PBq was added in 1992). A hot summer, followed by a dry winter in 1967, caused evaporation of the lake and dust from the shoreline containing around 20 TBq of ^{90}Sr and ^{137}Cs (in a 1 : 3 ratio) was blown over an area of 1800 km^2 and up to a distance of 75 km.[21] Up to four million m^3 of contaminated groundwater, containing in excess of 185 TBq has migrated 2.5 to 3 km away from Lake Karachay.[20]

As previously discussed, a waste storage facility became operational in 1953 and consisted of 20 stainless steel tanks utilising an external cooling system involving water flowing through a gap between the tank walls. The cooling system for one of these tanks failed and resulted in the evaporation, heat-up and ultimately explosion of the 70–80 tonnes of highly radioactive nitrate–acetate waste.[22] Around 740 PBq of activity was ejected from the explosion with approximately 90% falling out in the immediate vicinity of the accident and the remaining 10% forming a cloud extending to a height of 1 km. The fallout from this cloud exposed the Chelyabinsk, Sverdlovsk, and Tyumen regions to contamination.[22] In the immediate area, 1 to 2 km long by 0.5 to 1 km wide, the soil contamination amounted to 5180 TBq km^{-2} with contamination of the wider area, 75 km long and 7 km wide, amounting to around 1 TBq km^{-2}.[21]

2.3 United States of America

Like the United Kingdom, the United States has accumulated a nuclear legacy from over 60 years of research, production, use and storage of nuclear materials. Over this time frame, nuclear material was produced for use in both civilian power plants and in military weapons. The Department of Energy (DOE)'s 120 sites contain 40 million cubic meters of contaminated soil and debris and 1.7 trillion gallons of contaminated groundwater. Of this, at least 50% is contaminated with radionuclides, including caesium-137, plutonium-239, strontium-90, technetium-99, uranium-238, and uranium-235, as well as heavy metal contamination including chromium, lead and mercury.[23] The associated cleanup cost has been estimated to be in excess of a trillion dollars.[24] Contaminated sites include former uranium ore processing facilities such as Rifle, Colorado and Moab, Utah. The clean-up of these sites was tasked to the DOE under the Uranium Mill Tailings Remedial Action (UMTRA). During the years of operation of the Moab site, approximately 10.5 million tons of tailings and contaminated soils accumulated in an unlined pile 750 feet from the Colorado River.[25] In 2005, the DOE finalised the remediation strategy to be undertaken at Moab which included active groundwater remediation and off-site disposal of the tailings pile and other contaminated materials at the Crescent Junction disposal site.[26] The Rifle site consists of two old uranium processing plants: Rifle Old Processing Site and Rifle New Processing Site. Tailings and tailings-contaminated material from Rifle were transferred to the Rifle disposal site approximately six miles north of the Rifle New Processing Site and surface remediation was completed in October 1996. Contaminants of concern in the groundwater at both sites include arsenic, molybdenum, selenium, nitrate, uranium and vanadium, with contamination at New Rifle extending approximately three miles west of the site. Groundwater remediation is being achieved through natural flushing of the groundwater in conjunction with contaminant monitoring.[27] The *in situ* remediation of uranium was examined in a field scale study in 2003 in which acetate was injected into the subsurface over a three month period in order to stimulate microbial reduction of soluble U(VI) to insoluble U(IV) (ref. 28) and is discussed in detail later in this chapter. The Savannah River Nuclear Facility, South Carolina, was used to refine nuclear material for use in the United States defence program. The site used a system of canals and reservoirs to disperse heat from the reactors and consequently, various ponds connected to this system received cooling water discharges from the reactors. One such pond, Pond B, received discharges containing fission products such as ^{137}Cs, ^{90}Sr and ^{239}Pu. Radionuclide input peaked in 1963 and 1964, believed to be caused by the leakage of fuel elements stored in a water-filled basin in the reactor.[29] The vast majority of both ^{137}Cs (98%) and ^{90}Sr (85%) were found to be in the pond sediments.[29]

A number of other sites located in the United States have a more complicated environmental legacy left by the nuclear industry and are discussed below in more detail.

2.3.1 Rocky Flats

The Rocky Flats Environmental Technology Site, formerly Rocky Flats Nuclear Weapons Plant, is located northwest of Denver, Colorado, and between the years 1952 and 1989 was responsible for the production of components for the United States nuclear weapon program. This involved the use of various radioactive materials such as plutonium and uranium as well as toxic metals and hazardous solvents. Two main events are responsible for the release of plutonium outside the Rocky Flats Plant boundaries. These events were a fire that occurred in the plutonium processing building in 1957 and wind-blown releases occurring mainly during 1968 and 1969 from an outdoor waste storage area called the 903 Area. An estimated 5000 gallons of plutonium-contaminated waste leaked from the waste containers covering an area of 22 500 m^2 according to monitoring conducted in 1968. Windstorms in 1968 and 1969 blew the plutonium-contaminated soil off the site thus contaminating a much greater area, with an estimated 66.6 to 518 GBq of $^{239+240}Pu$ released to the off-site environment.[30]

Soil samples analysed from a series of pits along a contaminated toposequence at Rocky Flats revealed plutonium contamination ranging from 2220 to 11 460 Bq kg^{-1}, with a mean activity of 7250 Bq kg^{-1}, and ^{241}Am contamination ranging from 1840 to 8840 Bq kg^{-1}, with a mean activity of 5480 Bq kg^{-1}.[31] The activity was primarily located in the uppermost layer of the soil with 90% of the contaminants distributed in the top 20 cm in four of the five pits.

Synchrotron radiation studies revealed the oxidation state of the plutonium in the soils and concrete as Pu(IV) and identified its chemical form as the insoluble hydrous oxide $PuO_2 \cdot xH_2O$ (ref. 32) with transport of the plutonium likely confined to fine particle migration.

2.3.2 Oak Ridge

The Oak Ridge Reservation in east Tennessee consists of about 357 250 acres of land and contains the Y-12 Plant which played a historical role in the production of nuclear weapons.[33] Oak Ridge also houses one of the three DOE gaseous diffusion plants (K-25 Plant) used to enrich uranium. The Y-12 facility itself encompasses 324 hectares and is associated with mercury contamination arising from its use in nuclear weapons production through the 1950s and 1960s. An estimated 108 000 to 212 000 kg of mercury was released into the headwaters of the East Fork Poplar Creek over this time frame from the Y-12 plant.[34]

Further studies have showed that 77 180 kg of mercury are contained in the sediments and floodplain soils of a 15 mile stretch of the East Fork Poplar Creek which has its headwaters at Y-12. Some 227 kg are thought to leave this watershed annually.[35] The concentration of mercury in soils from the floodplains ranges up to 2400 $\mu g\ g^{-1}$ with an average of greater than 70% found to be metallic mercury and mercuric sulfide.[36]

Between the years 1951 to 1983, liquid acidic wastes (pH <2.0) containing metals (including uranium) dissolved in nitric acid were discharged into four seepage pits on the Y-12 complex known as the S-3 Ponds. The ponds were neutralised in 1983 by the addition of limestone, quicklime and sodium hydroxide until the pH reached greater than 9.0 resulting in the precipitation of calcium, iron and aluminium compounds.[37] Leakage from the S-3 ponds has created a contaminated groundwater plume in the underlying shale bedrock which extends more than 2 km both east and west of the ponds. Analysis of soils from this site has revealed sorbed and precipitated uranium concentrations of up to 800 mg kg^{-1}.[38] A summary of groundwater contaminants in the plume extending from the S-3 ponds is given in Table 3.

Uranium and technetium are likely to exist as the mobile U(VI) (as UO_2^{2+}) and Tc(VII) (as TcO_4^-) forms, although in the low pH conditions and in the presence of high nitrate and sulfate concentrations the uranium may be associated with nitrate (as $UO_2NO_3^+$) or sulfate (as UO_2SO_4 or $UO_2(SO_4)_2^{2-}$).[39] These conditions present problems for remediation strategies such as *in situ* bioremediation as the groundwater requires pre-treatment so that conditions are favourable for

Table 3 Contaminants in groundwater plume extending from S-3 ponds due to leakage of the ponds and dissolution of the shale and carbonate bedrocks due to acidic nature of the groundwater (pH ~3.5).[39]

Contaminant	*Maximum concentration*
U	0.2 mM
Tc	47 nM
Al	18 mM
NO_3^-	100 mM
SO_4^{2-}	100 mM
Ca^{2+}	25 mM
Mg^{2+}	8 mM
Co^{2+}	0.02 mM
Ni^{2+}	0.2 mM

microbially mediated reduction, discussed later in this chapter. High nitrate concentrations can potentially present further problems for remediation as in addition to being a competing electron donor to U(VI), it is also an effective oxidant of U(IV), leading to the oxidation and remobilisation of U(VI) through a variety of possible mechanisms.[40]

2.3.3 *Hanford*

The Hanford Nuclear Reservation, located on the Columbia River in Washington State, USA, was a plutonium production site which began operation in 1945. As a consequence of its former activities, a number of contamination issues have arisen. It has been estimated that more than 436 TBq of ^{239}Pu, 1065 TBq of ^{241}Am and 2 TBq of ^{237}Np were disposed of as liquid waste to the near surface at Hanford,[41] with 86% of the ^{239}Pu, 97% of the ^{241}Am and 77% of the ^{237}Np released into the Plutonium Finishing Plant (PFP) zone of the site. Despite the release of large quantities of radionuclides into the Hanford Vadose zone, only negligible amounts have entered the groundwater. Filtered samples from on site wells have so far shown no activity above the DOE-derived guide for ^{239}Pu (1.11 Bq l^{-1}) and only two unfiltered samples taken in 2006 exceeded this limit (1.3 and 1.5 Bq l^{-1}).[41]

Although the majority of plutonium on site is reported to be immobile, there are areas where vertical migration of plutonium and americium has occurred. The Z-9 trench is considered a "worst case" representation of the disposal area due to the acidic (pH 2.5), high salt waste solution containing nitrate (~5 M), aluminium (~0.6 M) and organic solvents. As much as 140 kg of plutonium was disposed of in this waste and although ~58 kg of plutonium was recovered in 1978, $^{239+240}$Pu has been found to be concentrated (up to 9.25 MBq kg^{-1}) in silt layers 15–20 m below the surface correlating with the occurrence of co-disposed tributylphosphate (TBP).[42] Americium-241 was also found to be accumulated at this point, and also concentrated at a second horizon at the bottom of the underlying fine-grained Cold Creek Unit (~40 m below ground surface) at levels greater than 11.1 MBq kg^{-1} with no accompanying TBP.[42] A number of possible conditions may have contributed to this vertical migration including the acidic nature of the waste, formation of soluble complexes and suspension and transport of colloids or nanoparticles.

In order to generate the ^{239}Pu needed to produce nuclear weapons, a very large quantity of uranium (either as metal or UO_2) was irradiated. The subsequent retrieval of plutonium from the matrix resulted in a large volume of aqueous waste containing high concentrations of uranium. This waste (also containing other fission products) is stored in 177 underground steel tanks in different areas of the Hanford site referred to as "Tank Farms" which are subdivided into Waste Management Areas. A large number (68 out of 149) are known or suspected to have leaked to date with the largest release occurring in 1951 in the 241-BX-Tank Farm. Nearly 3.5×10^5 litres of highly radioactive waste containing more than 7000 kg of uranium was released into the subsurface.[43] Between the years 1944 and 1988, almost two million cubic meters of tank waste was generated with subsequent evaporation, discharge, chemical treatment and leakage reducing this volume to 200 000 cubic meters. This makes up to 60% of the current tank waste which contains around 7.03×10^6 TBq of radioactivity and 170 000 tonnes of chemicals with each cubic meter of tank waste containing nearly 37 TBq of radioactivity.[44]

The mobility of uranium in the contaminated sub-surface beneath the tank farms has been shown to vary depending on the surface phases present at different depths. Uranium silicate precipitates were found in relatively shallow sediments, whereas the uranium was found adsorbed to sediment surfaces at intermediate and deeper depths, both in the form of U(VI).[45] Migration of uranium in the shallow sub-surface may therefore be slow as it relies on the slow process of mineral dissolution. In contrast, migration may be relatively fast in deeper conditions as surface desorption processes occur over a faster time frame. Work conducted on sediments taken from boreholes near to the storage tanks at Hanford revealed that the uranium is again predominantly found in the U(VI) state, with approximately 51% to 63% labile and therefore potentially mobile, with the remaining portion locked up in mobilization-resistant phases.[46]

A history of liquid waste disposal activities is provided by Gephart[44] and is briefly summarised here. Liquid waste has been dealt with *via* a number of methods during the operations of the Hanford site. In 1944, during fuel reprocessing, liquids which were only mildly contaminated were dumped into depressions on the ground, contaminating both the sandy sediments and eventually the groundwater. Some of these contaminants were blown downwind thus contaminating an even greater area. When dumping of these liquid wastes was halted they were instead pumped down reverse wells which lead to contaminants being injected closer to, or directly into, the underlying hydrology, bypassing the overlying sediment which could otherwise have acted, *via* sorption, as a sink. When this process was stopped after only a few months, liquid wastes were pumped directly into shallow buried box structures, gravel-filled tile fields, buried concrete pipes and open trenches later backfilled with gravel.

These processes, compounded by tank leakages, have led to the contamination of up to 28 300 m^3 of soil,[47] which along with the contaminated groundwater contains around 8325 TBq of ^{137}Cs, 6660 Tbq of ^{3}H, 1924 TBq of ^{90}Sr, 1850 TBq of Pu and 25.9 TBq of ^{99}Tc (ref. 44). Groundwater underlying around 12% of the Hanford site contains carbon tetrachloride, chromium, nitrate, ^{90}Sr, ^{99}Tc, ^{129}I and uranium at levels above the drinking water standard.[48] Although the site groundwater is not a source of public drinking water and does not significantly affect off-site water resources, contaminants such as ^{99}Tc and ^{129}I are mobile in groundwater and thus can migrate deep into the vadose zone and could potentially enter aquifers.

Earlier work estimated that up to 4×10^{16} Bq of ^{137}Cs had leaked into the vadose zone from the tank farms with measured activity from contaminated sediments as high as 10^5 Bq g^{-1}.[49] The waste stored in these tanks also typically contained a significant concentration of high ionic strength solutions including $NaNO_3$ (>0.5 mol l^{-1}).[49] In the presence of high salt concentrations, caesium was only found to absorb to high-affinity, frayed edge sites of mica minerals with sodium being an effective competitor for such sites.[50] The high sodium released in the leaked Hanford waste may therefore prevent the retardation of ^{137}Cs at the site. Borehole data also suggests that caesium is not undergoing significant sorption as peak ^{137}Cs activities were detected between 20 and 26 m, reaching up to around 40 m, beneath the SX tank farm responsible for the majority of the caesium release.[51]

3 Depleted Uranium

Contamination from depleted uranium (DU) is an issue covering several sites worldwide. The properties of DU, such as its high density (19.05 g cm^{-3}) and penetrating strength, have led to its use in a number of civil and military applications including munitions. Such munitions have been used in a number of conflicts over the past few decades with a summary provided in Table 4. Many of these rounds miss their target and can penetrate some distance into the ground.

The experimental test firing of depleted uranium munitions is also responsible for contamination at various firing ranges in both the UK and the USA. The UK Ministry of Defence (MOD) estimates that 15 tonnes of DU rounds were fired at an armour plate at the Eskmeals firing range in Cumbria between 1981 and 1995, with an additional 30 tonnes fired into the Solway Firth at Kirkcudbright, Scotland since 1982 (ref. 52). Experimental firing of DU rounds began in the USA at the Aberdeen and Yuma proving grounds in the early 1970s. More than 70 tonnes of DU have been deposited over 1500 acres at the Aberdeen Proving Ground, Maryland, into the sediments and the aquatic environment.[53,54]

Table 4 Overview of DU munitions fired in conflicts from the past few decades and subsequent contamination created.[55,56] (Adapted from A. Bleise, P. R. Danesi and W. Burkart, *J. Environ. Radioactiv.*, 2003, **64**, 2–3).

Conflict Zone	*Contamination*	*Source*
Iraq and Kuwait (1990–1991)	321 tonnes of DU	United States • Air Force fired 783 514 rounds of 30 mm DU ammunition • Army fired 9552 DU tank rounds United Kingdom • Less than one hundred 120 mm DU rounds
Bosnia-Herzegovina (1994–1995)	3 tonnes of DU	NATO airstrikes • About 10 800 DU rounds
Kosovo (1999)	10 tonnes of DU spread over 112 sites	A-10 antitank aircraft fired ~30 000 rounds (30 mm)
Iraq (2003)	2 tonnes known 170–1700 tonnes speculated	2 tonnes fired by the UK MOD Amount fired by USA forces not yet disclosed but speculative figures range from 170 to 1700 tonnes

Natural and depleted uranium share a similar chemotoxicity but the radiotoxicity is around 60% higher for the former. The low specific radioactivity combined with the dominance of alpha emissions means that no acute risk is associated with external exposure to DU but internal exposure presents serious health risks. Therefore, the main risk arises from DU dust generated from the impact of DU munitions on hard surfaces. Re-suspension of settled DU dust can occur if the particle size is sufficiently small. Traces of ^{236}U and $^{239+240}Pu$ have been found in DU penetrators collected in Kosovo[55] and trace amounts of americium, neptunium and ^{99}Tc are also thought to be present in DU.[56]

A review by the Royal Society[57] estimates that in a worst case scenario for DU exposure in the battlefield, a soldier who experiences level I exposure to DU (exposure dominated by inhalation of aerosols generated by DU impact) has a increased risk of 1.2 per 1000 of death from lung cancer. However, they cite that poor data collection on battlefield exposure makes estimating such health risks very difficult. DU fragments left on the battlefield also pose a concern as a slightly increased risk of skin cancer is expected from long-term exposure to DU penetrators. This is of particular concern for children who may be attracted to such objects. DU penetrators remaining in the ground also pose a longer term risk through potential migration to food sources or into water supplies. The mobility of the DU from the contaminated "hotspot" depends on a number of factors including corrosion rates, DU particle re-suspension, and proximity to surface soils and water sources. Although this form of radionuclide contamination has been the focus of much recent media interest, there is comparatively little work published on the scale of the problem, or strategies to decontaminate environments contaminated by DU munitions. However, the reader is directed to a recent review on the environmental fate of DU for a more detailed critique.[58]

4 Remediation

A number of different techniques are available for the remediation of both groundwater and soils and can be categorised into biological, chemical and physical treatments. This review will look to provide an overview of some of these key techniques and will then focus on a number of case studies where these methods have been applied in the field. The advantages and drawbacks of the major techniques are summarised in Table 5.

4.1 Bioremediation

Biological treatments, referred to as bioremediation, encompass several techniques which can involve the redox transformation, biological accumulation or breakdown of a contaminant. Chemical speciation (oxidation state and complex form) is one of the primary controls on the mobility of metal contaminants in the environment, affecting both their solubility and reactivity with surfaces. For example, the metal chromium is mobile and highly toxic in the Cr(VI) state, but is both less mobile and up to 1000 times less toxic as the Cr(III) oxidation state.[59] The radionuclide, ^{60}Co can form a stable and mobile complex with

Table 5 A summary of the main advantages and disadvantages of key remediation practices.

Technique	*Advantages*	*Disadvantages*
Biotransformation	Can be performed *in situ* and *ex situ*. Relatively low cost compared to physiochemical methods. Highly selective treatment of contaminants.	Potential for re-oxidation and re-mobilisation of metals and radionuclides. Complex groundwater or soil chemistry can complicate or prohibit treatment. Regular monitoring required to assess effectiveness. Can only operate in conditions required for cell growth (*i.e.* limited pH range).
Biosorption	Can be performed *in situ*. No additional nutrients required. Relatively low cost compared to physiochemical methods. Not governed by physiological constraints of living cells. No secondary waste produced. Metal recovery is possible, especially from process waters. Specific contaminants can be targeted.	Early saturation can require metals desorption to continue use. No potential for degradation of compounds. Targeting certain contaminants may require the cultivation and introduction of species not natively present. Very limited commercial application.
Bioaccumulation	Can be performed *in situ*. Relatively low cost compared to physiochemical methods. No secondary waste produced. Specific contaminants can be targeted *e.g.* Cs^+ transported by K^+-uptake processes.	Requires subsurface conditions favourable for microbial metabolism. Toxicological effect on cell may inhibit cell metabolism or lead to cell death. Targeting certain contaminants may require the cultivation and introduction of species not natively present. Very limited commercial application.
Biomineralisation	Relatively low cost. *In situ* technique. Metals can be immobilised in the subsurface so no further treatment of waste is required.	May only operate over a specific pH range in certain cases. Mineral precipitation may clog pore spaces restricting groundwater flow to contaminants further from injection wells.
Phytoremediation	Can be performed *in situ*. Cheaper than most other *ex situ* and *in situ*. Plants can be easily monitored to assess effectiveness. Recovery and re-use of valuable metals is possible. Non-environmentally disruptive.	Treatment is limited to the surface area and depth of the plant roots. Possibility of contaminants entering the food chain. Slow growth and low biomass require long-term commitment. Saturation of contaminants may lead to toxicity affecting plant survival.

Table 5 Continued.

Technique	*Advantages*	*Disadvantages*
Chemical Oxidation	Can be performed *in situ*. Rapid treatment time. Ability to treat high concentrations of contaminants.	Non-selective High capital and operating costs Most methods operate over a narrow pH range
Sediment Washing	Closed system allows easier control of geochemical conditions. Can treat both organic and inorganic contaminants in the same system. Relatively low cost.	*Ex situ* technique. Ineffective in removing metals in the residual phase of sediments. Certain chelating agents used present an environmental risk themselves.
Electrokinetic	Ability to treat organic and inorganic contaminants simultaneously. Can operate in zones of low hydraulic flow through induction of electric field. Effective at removing high concentrations of contaminants. Can operate *in situ*. Contaminants can be removed with electrodes.	Non-selective, problems can arise if target ions are in much lower concentrations than non-target ions. Corrosion of anodes in acidic conditions. Contaminants removed may require further disposal. Precipitation of metals close to electrode can impede process. Requires continued operational costs.
In situ Vitrification	Can treat organic, inorganic and radionuclide contaminants simultaneously. Can be completed *in situ* with fused glass blocks remaining in place. Compacts original volume of contamination by up to 20–50%. One-step, fast process. Helps prevent leaching of contaminants.	Water in soils affects operational time and costs. Requires special equipment and training. High energy input needed.
Permeable Reactive Barrier	Can be performed *in situ*. Ability to treat multiple contaminants simultaneously. Typically low capital and operating costs compared with pump and treat systems. Variety of reactive media can be used to target specific contaminants. Long-term efficiency can be improved through adsorption from secondary precipitated minerals. Passive system requiring no ongoing energy input.	Mineral precipitation may passivate certain reactive media. Groundwater flow must be well characterised. Mineral precipitation may reduce permeability of barrier and affect groundwater flow. Limited to shallow depths (<15.24 metres) due to construction challenges.

ethylenediaminetetraacetic acid (EDTA) in the Co(III) state, but is less stable and hence less mobile in the Co(II) state.[60]

It has long been established that microorganisms are able to reduce metals,[61,62] with more recent work showing they are able to use such processes to conserve energy for growth. Focussing on reductive transformations, microbes are able to use some metals as the terminal electron acceptor during anaerobic respiration, in environments where oxygen has been depleted. Thus, stimulating their activity in the subsurface can cause the reduction of high oxidation state metal contaminants to less soluble forms and hence retard their migration. The mechanisms involved in microbial metal and radionuclide reduction are described in detail elsewhere.[59,63] Microorganisms are also able to reduce and degrade some organic contaminants through analogous respiratory processes when supplied with a suitable electron donor. For example, almost 98% of tetrachloroethylene (PCE) underwent complete reductive dechlorination to ethane when a laboratory column experiment used Rhine river sediment supplied with lactate as an electron donor.[64] Trichloroethylene (TCE), an industrial solvent and common subsurface contaminant,[65] was also shown to be degraded by the methanotrophic bacterium *Methylosinus trichosporium* OB3b in a co-metabolic process in a copper deficient medium.[66] The reader is directed towards a recent review by Pant and Pant, for a detailed account on the microbial remediation of TCE.[67]

Metal and radionuclide transport can also be restricted through precipitation with enzymatically generated ligands, such as sulfide[68,69] and phosphate (see Figure 1).[63] If supplied with an excess of these ligands then most of the metal should be removed from solution. An advantage to this method is that high concentrations of ligand are generated close to the cell surface which can act as nucleation foci for the onset of metal precipitation. An integrated approach to metal remediation using sulfur-cycling bacteria has been demonstrated.[70]

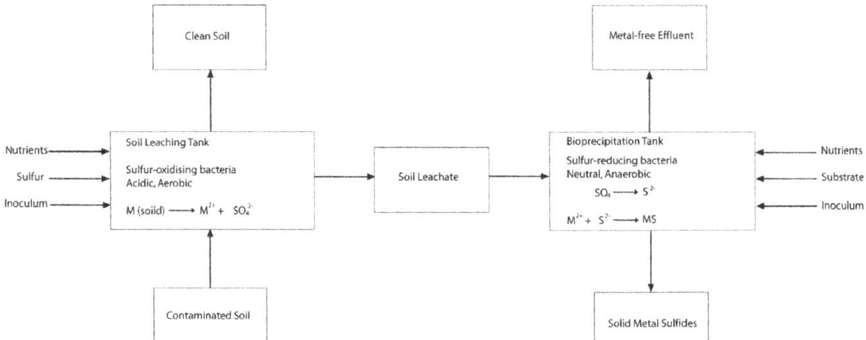

Figure 1 Diagram illustrating an integrated approach to bioremediation of metal-contaminated soils. The conditions and inputs required for bioleaching and bioprecipitation are displayed along with the outline reactions for each stage where M^{2+} = target metal ions (considered as divalent cations). (Adapted from C. White, J. A. Sayer and G. M. Gadd, *FEMS Microbiol. Rev.*, 1997, **20**, 3–4).

In this study, a number of metals were leached from artificially contaminated soil through the production of sulfuric acid by sulfur-oxidising bacteria. This leachate was then applied to a bioreactor containing sulfate-reducing organisms where greater than 80% of the metals were precipitated as solid metal sulfides.

The bacterial strains *Rahnella* sp. and *Bacillus* sp. were both shown to be capable of hydrolysing sufficient organophosphate to remove up to 95% of uranium in a simulated groundwater system. The system was most efficient between pH 5.0 and 7.0 with EXAFS spectroscopy identifying the uranyl phosphate precipitate as an autunite/meta-autunite group mineral.[71] This builds on earlier work on a *Citrobacter* (now classified as a *Serratia*) strain which coupled the efflux of phosphate driven by phosphotase-mediated breakdown of glycerol-2-phosphate to efficient uranium precipitation.[72] A case study involving phosphate biomineralisation at the Hanford site is discussed in detail later in this chapter. The biosorption and bioaccumulation of metals may act as a component in metal remediation through sorption of metals to cell surfaces or uptake into the cell. This can occur as a physiochemical, metabolic-independent mechanisms whereby metals sorb onto the surface of biomass or *via* metabolic-dependent processes in which the metal is taken up into the cell where it may precipitate locally and accumulate. Both processes have been reviewed extensively but a lack of commercial development has weakened continued research into this field.[73–76]

These techniques can be achieved through several different methods. Biostimulation involves the addition of key nutrients, such as an electron donor and carbon source, to the subsurface to stimulate the native microorganisms, usually done *via* injection wells. Advantages of such a method include the stimulation of extant bacteria that are already well suited to the environmental conditions and distributed throughout the subsurface. Relying on the local geology and hydrogeology to distribute the nutrients evenly can however, prove to be a disadvantage.

If the native bacteria do not have the metabolic capability to remediate a particular contaminant then bioaugmentation can be employed where by specialised microorganisms are added to the subsurface, along with the required nutrients, in order to remediate the contaminant. A number of reviews are available on the processes involved in bioaugmentation.[77,78] Both the aforementioned techniques operate *in situ* but *ex situ* bioremediation is also a possibility. *Ex situ* treatment involves the excavation of contaminated soil or pumping of groundwater into an above ground facility where the biological conditions can be better controlled. Although excavation and pumping is more expensive than *in situ* treatments, benefits include being able to adjust to aerobic or anaerobic conditions as required. The ability to operate in aerobic conditions allows certain bacteria to utilise organic contaminants, such as petroleum hydrocarbon mixtures and polycyclic aromatic hydrocarbons, as their source of carbon and energy thus potentially degrading the contaminants completely to CO_2 and H_2O. A further advantage of *ex situ* remediation is the ability to homogenise and continuously monitor the soil to ensure complete treatment occurs. Numerous studies examining the effectiveness of *ex situ* bioremediation have been performed.[79–81]

Phytoremediation, which utilises the ability of plants to degrade or accumulate contaminants, can also be employed in the remediation of soil and groundwater. The cost-effectiveness and non-environmentally disruptive nature of phytoremediation offers advantages over other bioremediation techniques. Further advantages include the ability to easily monitor the plants and the possibility of recovering and re-using valuable, accumulated metals. However, there are a number of disadvantages associated with this process which includes remediation being limited to the surface area and depth of the plant roots, the possibility of contaminants entering the food chain and the usually long period of time phytoremediation requires for completion. For further details, the reader is directed to a number of recent reviews.[82,83]

4.2 Chemical Redox Reactions

This technique is based on the *in situ* delivery of chemical oxidants to the contaminated media to destroy the contaminants by converting them to harmless compounds. Typical oxidants applied in this process include hydrogen peroxide, potassium permanganate, ozone and dissolved oxygen. A common application of this procedure, based on Fenton's Reagent, involves the addition of hydrogen peroxide and an iron catalyst to the contaminated area, generating a hydroxyl free radical:

$$H_2O_2 + Fe^{2+} => Fe^{3+} + OH^- + OH^* \qquad (1)$$

This free radical is capable of oxidising complex organic compounds, such as TCE, PCE, dichloroethylene (DCE), benzene, polyaromatic hydrocarbons (PAHs) and polychlorinated biphenyl (PCBs), with any residual hydrogen peroxide decomposing into water in the subsurface. Fenton's Reagent oxidation is most effective in acidic environments (pH 2–4) and becomes ineffective under moderate to strongly alkaline conditions.[84] Ozone can oxidise contaminants directly or through the production of hydroxyl radicals and is also most effective in acidic conditions. Due to its instability and high reactivity, ozone is produced onsite and requires injection *via* closely spaced delivery points. Permanganate is typically provided as a liquid or as solid potassium permanganate ($KMnO_4$) but is also available in sodium, calcium or magnesium salts. Permanganate reactions occur at a slower rate compared to ozone and peroxide and, depending on the pH, can destroy the contaminant through direct electron transfer or free radical oxidation. An advantage of permanganate use includes an operable pH range of 3.5 to 12.[84]

This method was used in a field study at the A/M Area of the Savannah River Site where undissolved dense non-aqueous phase liquid (DNAPL), including TCE and PCE, contamination was present. The treatment test operated over six days and used hydrogen peroxide and iron sulfate to generate hydroxyl radicals to destroy approximately 600 pounds of DNAPL-contaminated soil in the target area. After the trial period, 94% of the targeted DNAPL was destroyed[85] at a total cost of \$511k for the project demonstration.

4.3 Permeable Reactive Barrier

The use of a Permeable Reactive Barrier (PRB) involves placement in the subsurface of a barrier consisting of a permanent, semi permanent or replaceable reactive media across the flow path of a contaminated groundwater plume. As the groundwater passes through the barrier under its natural gradient, contaminants are either degraded by, or retained in, the reactive media in a passive treatment system. A typical PRB involves the excavation and backfilling of a continuous trench with a reactive material designed to target particular contaminants. Examples of reactive media used include iron, limestone, calcium phosphate-based minerals, compost and activated carbon, with iron being the most common.[86] A review of the uses of these various reactive media is provided by Thiruvenkatachari *et al.*[87] Zero-valent iron (ZVI) acts as a reactive medium through corrosion/oxidation of the metal *in situ* and donation of electrons from this process to organic and inorganic contaminants, such as halogenated hydrocarbons, U(VI) and Cr(VI), which are reduced thereby leading to degradation of the organic contaminant or metal immobilization.[87] Consequently, the long-term efficiency of ZVI barriers is heavily dependent on the corrosion of Fe^0, as continued use results in authigenic mineral formation which restricts the availability of reactive Fe^0.[88] However, the precipitation of ferrihydrite clusters found away from the immediate surface of the Fe^0 barrier provides an increase in potential sites for metal adsorption thus prolonging the life of the PRB.[88,89] A PRB containing ZVI was used in a remediation effort at Oak Ridge and is discussed in more detail later in this chapter.

Advantages of using a PRB include the *in situ* capture of contaminants, alleviating the need to manage the waste generated by pump and treat methods. Additionally, multiple contaminants, such as metals, radionuclides and organics, can be treated simultaneously;[90] and both operating and maintenance costs are typically low.[91] A review of the long-term performance of PRBs is presented by Henderson and Demond.[92]

4.4 Sediment Washing

Sediment washing is a relatively simple, typically *ex situ* technique involving the cleaning of contaminated soils with various reagents. Depending on the nature of the contaminant, a number of additives can be employed in the washing process including acid washing (*e.g.* H_2SO_4 and HNO_3) and chelating agents [*e.g.* EDTA, diethylenetriaminepentaacetic acid (DTPA) and ethylenediamine-*N*,*N′*-disuccinic acid (EDDS)] to assist with the solubilisation and desorption of the metal from the sediment. This technique is useful for weaker bound metals, those associated with the exchangeable, carbonate and reducible oxide fractions of the soil, but is inefficient at removing metals in the residual fraction.[93] Acid washing can be applied through a variety of abiotic and biological *ex situ* techniques.[70,94] Chelating agents can be used in soil washing to remove contaminants from sediments through the formation of stable metal chelate complexes which can then be removed in solution.[95] EDTA has been studied extensively for use as a chelating agent for use in soil washing[96,97] and can enhance metal mobilisation *via* two mechanisms: fast thermodynamically favourable complexation between EDTA and certain cationic metals, as well as slow driven EDTA-dissolution. The former involves the breakdown of some weak soil-metal bonds while the latter can partially disrupt the soil structure thus mobilising metals bound to oxides and organics.[98]

4.5 Electrokinetic Remediation

Contaminants are treated by electrokinetic remediation through the application of a low intensity electric current between a cathode and an anode placed within the contaminated soil. Through this process, organic, inorganic and radioactive contaminants can be separated and extracted from clay-rich soils, sludges and sediments. Application of the electric field creates an acidic front around the anode, due to an excess of H^+ ions, and an alkaline front at the cathode, due to an excess of OH^- ions. The electric gradient created initiates the movement of water, charged chemicals and charged particles through the processes of electro-osmosis, electromigration and electrophoresis, respectively, moving anions towards the positive electrode and cations towards the negative. The contaminants can then be removed through electroplating or precipitation at the electrodes, the use of ion exchange resins or pumping the waste to the surface.[99] Complexing agents, surfactants and other reagents can be used to increase the efficiency of treatment.[100]

A paper by Lageman examines the processes involved in electrokinetic remediation and examines numerous sites where inorganic and organic contaminants have been treated through this technique.[101] Work by Cundy examines the use of electrokinetics to generate a Ferric Iron Remediation and Stabilisation (FIRS) barrier. By applying a low, direct electric potential between two or more sacrificial Fe-rich electrodes placed in the contaminated soil, a strong pH/E_h gradient can be generated in the soil column. This forces the precipitation of an Fe-rich barrier between the electrodes.[102] Soil samples were taken from the Ravenglass estuary, Cumbria, UK, containing artificial radionuclides from the nearby Sellafield plant and placed into a Perspex cell. Cast iron electrodes were embedded into the soil and a potential of 1.5 V was applied between them for 17 days. After this time, a 30% reduction in ^{60}Co was observed in the anode zone with a 50% enrichment in the iron band. Manganese, calcium and strontium were also depleted in the anode zone and enriched on, or around, the iron band. Arsenic, which was desorbed at the high pH found in the cathode zone, was found to be 100% enriched in the iron band. The radionuclides, plutonium and americium, were not found to be significantly mobilised over the this time frame.[102] The use of electrodes as a potential electron donor for microorganisms resulted in the removal of U(VI) from solution in a study by Gregory and Lovley. When the electrodes were poised at −500 mV in the absence of microorganisms, U(VI) was removed from solution but was returned to solution when the poise at the electrodes was removed. If *Geobacter sulfurreducens* was present on the electrode, then U(VI) did not return to solution suggesting the uranium was reduced from U(VI) to

U(IV).[103] A review on the electrical stimulation of microbes is provided by Thrash and Coates.[104]

Electrokinetic remediation offers advantages as a treatment method through: the ability to treat both inorganic and organic contaminants at the same time; being able to treat contaminants in areas of low hydraulic flow by inducing movement of water, ions and colloids through an electric field; and competitive cost and effectiveness. However, the process can be ineffective when target ions are in much lower concentrations than non-target ions, and corrosion of the anodes in acidic conditions presents *in situ* treatment problems.

5 Case Studies

A number of the remediation methods described above have been utilised in various field tests at sites suffering with radionuclide contamination. Three of these case studies are discussed below.

5.1 Hanford Case Study

Bench- and field-scale studies were performed by the Pacific Northwest National Laboratory along with a number of collaborators in order to test the remediation potential of using polyphosphate injections to reduce uranium concentrations in groundwater beneath the contaminated 300 Area of the Hanford Site. A detailed study is provided in a number of PNNL reports[105–107] and will be summarised briefly in this section. The concept of polyphosphate injections works by the formation of stable and insoluble uranium phosphate minerals (autunite) and phosphate precipitates (apatite) for uranium sorption.[71,108] As autunite sequesters uranium as U(VI) rather than reducing it to U(IV), the issue of re-oxidation and consequent remobilisation is nullified offering a potential advantage over bioreduction methods. Phosphate minerals precipitate when phosphate-containing compounds degrade in water, due to hydrolysis, and hence rapid mineral formation can occur in an aquifer resulting in a reduction in permeability. However, the longer the phosphate chain, the slower the hydrolysis and consequently the use of long-chain polyphosphate compounds results in a lower change in hydraulic conductivity.[109]

The test site chosen for the field scale study was located in the 300 Area of the site and involved a three-stage approach to the polyphosphate injections. Water was routed from an extraction well located 190 m from the injection well. Sampling pumps were installed in all site monitoring wells, capable of delivering flows up to 7.57 litres per minute. The sample tubing from these wells was routed directly into a mobile laboratory and connected to a sampling manifold which monitored field parameters (Eh, pH, temperature and dissolved oxygen) and collected samples for anion, cation and trace metal analysis.

Based on previous laboratory studies summarised by Vermeul,[105] a three-phase injection strategy was identified in order to generate both the uranium-bearing autunite and uranium sorping apatite. An initial injection of polyphosphate was delivered to the subsurface to initiate the formation of autunite, followed directly by an injection of calcium chloride to allow the formation of calcium phosphate, apatite. The process was concluded with a final injection of polyphosphate following on directly from the $CaCl_2$ injection. A contribution of 25% orthophosphate, 25% pyrophosphate, and 50% tripolyphosphate made up the phosphorous in each polyphosphate injection.

Formation of apatite is affected by the mixing time between the polyphosphate and calcium species which proved to be variable throughout the site. Phosphate data indicated that wells in a radial distance of 23 m from the injection site received between 40% and 60% of the injection concentration. This suggests that a relatively large lateral area could be treated *via* the formation of autunite although apatite formation may be problematic.

Uranium concentrations in the targeted treatment zone were typically between 60 and 80 $\mu g\, l^{-1}$ prior to the injections. Uranium monitoring data from wells inside the target area showed an initial decrease in the concentration of uranium, to below the drinking water standard of 30 $\mu g\, l^{-1}$, but a significant rebound was observed about two months after the treatment. At a well outside the treatment zone, uranium concentrations were not observed to decrease until one month after treatment and then displayed a slower rebound. This suggests that uranium concentrations were effectively decreased through the formation of uranyl-phosphate mineral phases (autunite) and were then recharged by the uranium plume on site. However, it is also possible that the uranium decrease was partially due to displacement by the injections of large volumes of high ionic strength solutions. It was thought that long-term remediation would occur *via* the sequestration of uranium through adsorption to apatite and subsequent conversion to stable uranyl-phosphates (autunite) but these data suggest that this is not the case at this site.

5.2 *Rifle Case Study*

As discussed previously, the Rifle UMTRA site in Colorado is an old uranium processing facility which suffers from various contamination issues, including uranium. Uranium is predominantly found in the mobile U(VI) form in the subsurface due to an insufficient supply of electron donors to stimulate anaerobic respiration and/or consume dissolved oxygen. Laboratory studies have demonstrated the potential of microbes to reduce U(VI) to immobile U(IV) in an aquifer system,[110] and the *in situ* treatment of U(VI) using the same method was tested at the Old Rifle site. Contaminated soil has been removed from the site, leaving only groundwater contamination within the local aquifer. Concentrations of uranium in this area range from 0.4 to 1.4 μM, above the maximum UMTRA contamination limit of 0.18 μM.[28]

The method used in this field-scale test was describe in detail by Anderson *et al.*,[28] and is summarised briefly here. Injection wells were installed in two rows of ten, perpendicular to groundwater flow (which is typically towards the Colorado River). Each well contained three injection points positioned at different depths in the subsurface. A storage tank was filled periodically with native groundwater and was amended with sodium acetate as an electron donor to stimulate uranium-reducing bacteria and potassium bromide as a conservative tracer at concentrations of 100 and 10 mM, respectively. Oxygen was removed from the groundwater through nitrogen sparging. During operations, the injections were set to provide 1 to 3 ml of the solution from the storage tank per minute corresponding to 1 to 3 mM acetate and 100 to 300 μM bromide per day. Monitoring wells were installed at intervals downgradient corresponding to groundwater travel of approximately 4, 9, and 18 days with a further three wells placed upgradient to serve as controls. Acetate was injected continuously over a three month period from June to October 2002, with groundwater samples collected at regular time intervals from all monitoring wells. Groundwater conditions were monitored, including pH, conductivity, redox potential and dissolved oxygen, with further samples taken for U(VI), anion (bromide, nitrate, and sulfate), Fe(II), sulfide and acetate analysis. A second round of acetate injections were made over the same months in 2003, after which no further amendments were made.[111]

Bromide, added as a groundwater tracer, was not detected in any of the upgradient wells but was detected after 4, 9, and 18 days at each of the corresponding downgradient wells confirming the injection solution had reached the targeted area. After the first set of injections, U(VI) concentrations were observed to decrease 9 days after the injections began with concentrations dropping to or below 0.18 μM within 50 days at some wells.[28] The decrease in U(VI) was concurrent with the accumulation of Fe(II) and prior to any sulfate reduction. After 50 days, the U(VI) concentration began to increase, coincident with a decrease of Fe(II) and acetate falling to non-detectable levels. Bromide levels were still detected at wells where acetate levels had fallen suggesting that an increase in consumption of acetate was occurring near the point of injection. This correlated with observations following the second injection stage of a depletion of reducible iron oxide near the injection point and an accumulation of sulfide[111] suggesting all the available Fe(III) had been consumed and that sulfate reducers were now actively consuming acetate at the injection point.

A substantial shift in the microbial community was observed throughout the injection trials. Organisms in the family *Geobacteraceae* (which includes the known U(VI)-reducing genus *Geobacter*) became dominant early on,[28] with the greatest enrichment of *Geobacteraceae* correlated to the greatest proportion of U(IV) detected.[111] As reducible Fe(III) became depleted and sulfide accumulation occurred, the dominance of the *Geobacteraceae* decreased as they were replaced by species related to known sulfate-reducers.[111] After the second round of injections in 2003, U(VI) continued to be removed from the groundwater for over a year after the cessation of acetate injections.[112] This casts doubt on the suggestion following on from the first round of injections that U(VI) removal is acetate dependent. Flow-through column experiments suggested that the continued decrease in groundwater U(VI) levels could be linked to increased sorption to soils in a reduced environment.[112]

This series of field studies suggest that the stimulation of metal-reducing bacteria is an effective method for the removal of U(VI) from groundwater. However, when the supply of reducible Fe(III) oxides runs out, sulfate-reducers become dominant and do not appear to be as effective at reducing U(VI) to U(IV). Promising data from the second round of injections indicates that in sufficiently reduced soils, U(VI) removal may continue, without the continued need for acetate injections, *via* sorption to soils.

5.3 Oak Ridge Case Study

As was discussed previously, a number of contamination issues exist at the Y-12 complex at Oak Ridge. In order to address these concerns, two permeable iron reactive barriers were installed at the Y-12 plant, consisting of two pathways, in 1997.[113] As water flows through the barrier, the reactive medium (in this case Fe^0) traps or degrades the contaminant. Pathway 1 at the Y-12 complex was designed to capture groundwater in a gravel-filled, high-density-polyethylene-lined trench. The groundwater would then be treated within a vault containing zero-valent iron. The second pathway involved a permeable trench, in a sub-parallel direction to groundwater flow. The trench, 2 ft wide and 225 ft long, contained a 26 ft long zone of ZVI covered either side by gravel backfilled zones. Groundwater samples from monitoring wells both in, and downgradient, of the iron barrier at Pathway 2 contained only very low concentrations of uranium ($<0.05\,mg\,l^{-1}$) compared to values found in groundwater samples in upgradient wells. This would suggest that ZVI is effective at immobilising uranium present in groundwater.[114] Uranium concentrations in middle and deep wells located within the iron barrier displayed slightly higher than expected levels of uranium at ~ 0.2 to $1\,mg\,l^{-1}$. These wells are located in the upgradient portion of the iron barrier where upward hydraulic gradients dominate. The higher concentrations seen here may therefore be a result of a higher inflow of untreated groundwater. Some downgradient wells also showed higher than expected uranium concentrations suggesting that treated groundwater is being re-contaminated from the mobilisation of uranium on downgradient soils or that groundwater flows not treated by the barrier are reaching the wells.[114]

Another contrasting field research study was performed using reduction of U(VI) to U(IV) as a method for immobilising the contaminant. Subsurface conditions favourable for bioremediation were established[38] followed by periodic injection of ethanol. The pH of the test area was adjusted to pH 5–6 causing an increase in uranium sorption and resultant decrease in groundwater uranium concentrations from ~ 300 (at an initial pH of ~ 3.4) to $\sim 5\,\mu M$.[115] Ethanol injections began on day 137 and ended on day 535. Following on from an initial denitrification phase (day 137–184), a period of uranium and sulfate reduction occurred (day 184–535) during which uranium concentrations in the groundwater decreased from 5 to $1\,\mu M$.[115] XANES analysis confirmed that between 39% and 53% of the uranium recovered from the sediments after biostimulation was reduced U(IV).[115] The results from this study, where U(VI) reduction correlated with sulfate reduction, contrast to those from a similar study at the Rifle Site (discussed previously), where aqueous uranium concentrations rebounded when sulfate reducing conditions became dominant.[28] It is possible that sulfate-reducing bacteria that are capable of U(VI) reduction were stimulated by the use of an electron donor (ethanol) in this study, in contrast to the use of acetate at the Rifle site which may have stimulated alternative organisms. Further studies at this site are required to assess the long-term resistance of U(IV) to re-oxidation and remobilisation.

A recent column-flow experiment demonstrated the potential for the remediation of uranium and technetium in low pH, highly contaminated environments, such as Oak Ridge (as discussed previously) through co-precipitation. Luo *et al.* showed that in conditions found at Oak Ridge, greater than 95% of soluble uranium and 83% of technetium can be co-precipitated with Al-oxyhydroxides by raising the pH above 4.5 with the addition of a strong base (NaOH).[39] The precipitated uranium and technetium were found to be stable in the presence of high nitrate concentrations [50 mM $Ca(NO_3)_2$] and low carbonate concentrations.[39]

6 Conclusions

The successful management of land contaminated by our global nuclear legacy is a major challenge that relies on the successful cooperation across disciplines including chemistry, biology and engineering. The scale and complexity of the problem is apparent from this short review and illustrated by recent estimated costs of $1 trillion for cleanup of legacy wastes in the USA.[23] Cleanup costs in the UK are of the order of £100 billion with the discounted cost recently put at £44.5 billion.[3] However, there is a range of flexible remediation options that have already been trialled, with some success, on land contaminated with radioactive waste, and have the potential to help treat the wide range of highly complex contamination scenarios that are present on nuclear sites worldwide. This is a matter of importance, as widespread public acceptance of nuclear power as an attractive energy source can only be achieved if the successful management of our nuclear legacy wastes can be demonstrated.

Acknowledgements

RK acknowledges funding from EPSRC and CASE support from AWE. JRL acknowledges financial support from the Royal Society, NERC, EPSRC and the BBSRC.

References

1. K. A. Rogers, *Prog. Nucl. Energy*, 2009, **51**, 2.
2. T. D. Bergan, *J. Environ. Radioactiv.*, 2002, **60**, 1–2.
3. Nuclear Decommissioning Authority, *Draft strategy published September 2010 for consultation*, Nuclear Decommissioning Authority, Moor Row, UK, 2010.
4. D. McCubbin, K. S. Leonard, R. C. Greenwood and B. R. Taylor, *Sci. Total Environ.*, 2004, **332**, 1–3.
5. B. Allard, U. Olofsson and B. Torstenfelt, *Inorg. Chim. a-F-Block*, 1984, **94**, 4.
6. J. Gray, S. R. Jones and A. D. Smith, *J. Radiol. Prot.*, 1995, **15**, 2.
7. A. B. MacKenzie, G. T. Cook and P. McDonald, *J. Environ. Radioactiv.*, 1999, **44**, 2–3.

8. Nuclear Decommissioning Authority, *Strategic Environmental Assessment: Site Specific Baseline, Capenhurst*, Nuclear Decommissioning Authority, Moor Row, UK, 2010.
9. DEFRA BERR and Devolved Administrations for Wales and Northern Ireland, *Managing radioactive waste safely: A framework for implementing geological disposal*, D.f.B. D.f.B. Department for Environment Food and Rural Affairs, Enterprise and Regulatory Reform and the Devolved Administrations for Wales and Northern Ireland, London, UK, 2008.
10. J. Hunter, *SCLS Phase 1 – Conceptual Model of Contamination Below Ground at Sellafield*, Nuclear Sciences and Technology Services, BNFL, UK, 2004.
11. Sellafield Ltd, *Groundwater Annual Report 2009*, Characterisation, Land Quality & Clearance, Sellafield Ltd, Sellafield, UK, 2009.
12. UKAEA, *Dounreay 'Interim' Integrated Waste Strategy*, Waste Services Unit, UKAEA, Abingdon, UK, 2008.
13. Dounreay Particles Advisory Group, *Dounreay Particles Advisory Group: Third Report*, Scottish Environment Protection Agency, Stirling, UK, 2006.
14. Dounreay Site Restoration Ltd, *Strategy for Contaminated Land Management and Restoration 2008*, Dounreay Site Restoration Ltd, Thurso, UK, 2009.
15. The Scottish Parliament, *Dounreay*, The Information Centre, The Scottish Parliament, Edinburgh, UK, 2001.
16. K. Suokko and D. Reicher, *Environ. Sci. Technol.*, 1993, **27**, 4.
17. UNSCEAR, *United Nations Scientific Committee on the Effects of Atomic Radiation, Exposures from the Chernobyl accident*, United Nations, New York, USA, 1988.
18. UNSCEAR, *United Nations Scientific Committee on the Effects of Atomic Radiation, Exposures and effects of the Chernobyl accident*, United Nations, New York, USA, 2000.
19. International Atomic Energy Agency, *Environmental Consequences of the Chernobyl Accident and their Remediation: Twenty Years of Experience*, IAEA, Vienna, Austria, 2006.
20. T. B. Cochran, R. S. Norris and K. L. Suokko, *Annu. Rev. Energ. Env.*, 1993, **18**.
21. B. F. Myasoedov and E. G. Drozhko, *J. Alloys Compd.*, 1998, **271**.
22. B. V. Nikipelov, G. N. Romanov, L. A. Buldakov, N. S. Babaev, Y. B. Kholina and E. I. Mikerin, *Sov. Atom. Energy*, 1989, **67**, 2.
23. J. McCullough, T. Hazen and S. Benson, *Bioremediation of metals and radionuclides: What it is and how it works*, Lawrence Berkeley National Laboratory, Berkeley, CA, USA, 1999.
24. A. Palmisano and T. Hazen, *Bioremediation of metals and radionuclides: What it is and how it works (2nd Edition) Lawrence Berkeley National Laboratory*, Berkeley, CA, USA, 2003.
25. B. J. Merkel, A. Hasche-Berger, K. Karp and D. Metzler, in *Uranium in the Environment*, Springer, Berlin and Heidelberg, Germany, 2006, 671.
26. Moab Fact Sheet. *Overview of Moab UMTRA Project Fact Sheet, U.S. Department of Energy*. http://www.giem.energy.gov/moab/documents/factsheets/20080801OVERVIEW.pdf (accessed 26th October 2010).

27. Rifle Fact Sheet, *Rifle, Colorado, Processing Sites and Disposal Site Fact Sheet, U.S. Department of Energy, Office of Legacy Management*, 2009.
28. R. T. Anderson, H. A. Vrionis, I. Ortiz-Bernad, C. T. Resch, P. E. Long, R. Dayvault, K. Karp, S. Marutzky, D. R. Metzler, A. Peacock, D. C. White, M. Lowe and D. R. Lovley, *Appl. Environ. Microbiol.*, 2003, **69**, 10.
29. F. W. Whicker, J. E. Pinder, J. W. Bowling, J. J. Alberts and I. L. Brisbin, *Ecol. Monogr.*, 1990, **60**, 4.
30. T. D. Margulies, N. D. Schonbeck, N. C. Morin-Voilleque, K. A. James and J. M. LaVelle, *J. Environ. Radioactiv.*, 2004, **75**, 2.
31. M. I. Litaor, G. Barth, E. M. Zika, G. Litus, J. Moffitt and H. Daniels, *J. Environ. Radioactiv.*, 1998, **38**, 1.
32. D. L. Clark, D. R. Janecky and L. J. Lane, *Phys. Today*, 2006, **59**, 9.
33. W. D. Bostick and W. H. Hermes, *Waste Manage.*, 1993, **13**, 5–7.
34. G. L. Liu, J. Cabrera, M. Allen and Y. Cai, *Sci. Total Environ.*, 2006, **369**, 1–3.
35. F. X. Han, Y. Su, D. L. Monts, C. A. Waggoner and M. J. Plodinec, *Sci. Total Environ.*, 2006, **368**, 2–3.
36. M. O. Barnett, L. A. Harris, R. R. Turner, T. J. Henson, R. E. Melton and R. J. Stevenson, *Water, Air, Soil Pollut.*, 1995, **80**, 1.
37. L. A. Shevenell, G. K. Moore and R. B. Dreier, *Ground Water Monit. R.*, 1994, **14**, 2.
38. W. M. Wu, J. Carley, M. Fienen, T. Mehlhorn, K. Lowe, J. Nyman, J. Luo, M. E. Gentile, R. Rajan, D. Wagner, R. F. Hickey, B. H. Gu, D. Watson, O. A. Cirpka, P. K. Kitanidis, P. M. Jardine and C. S. Criddle, *Environ. Sci. Technol.*, 2006, **40**, 12.
39. W. S. Luo, S. D. Kelly, K. M. Kemner, D. Watson, J. Z. Zhou, P. M. Jardine and B. H. Gu, *Environ. Sci. Technol.*, 2009, **43**, 19.
40. J. D. Istok, J. M. Senko, L. R. Krumholz, D. Watson, M. A. Bogle, A. Peacock, Y. J. Chang and D. C. White, *Environ. Sci. Technol.*, 2003, **38**, 2.
41. K. J. Cantrell, *Transuranic Contamination in Sediment and Groundwater at the U.S. DOE Hanford Site*, 2009.
42. A. R. Felmy, K. J. Cantrell and S. D. Conradson, *Phys. Chem. Earth*, 2010, **35**, 6–8.
43. J. P. McKinley, J. M. Zachara, C. X. Liu, S. C. Heald, B. I. Prenitzer and B. W. Kempshall, *Geochim. Cosmochim. Acta*, 2006, **70**, 8.
44. R. E. Gephart, *Phys. Chem. Earth*, 2010, **35**, 6–8.
45. W. Um, Z. M. Wang, R. J. Serne, B. D. Williams, C. F. Brown, C. J. Dodge and A. J. Francis, *Environ. Sci. Technol.*, 2009, **43**, 12.
46. W. Um, J. P. Icenhower, C. F. Brown, R. J. Serne, Z. M. Wang, C. J. Dodge and A. J. Francis, *Geochim. Cosmochim. Acta*, 2010, **74**, 4.
47. G. W. Gee, M. Oostrom, M. D. Freshley, M. L. Rockhold and J. M. Zachara, *Vadose Zone J.*, 2007, **6**, 4.
48. T. M. Poston, J. P. Duncan and R. L. Dirkes, *Hanford Site Environmental Report for Calendar Year 2008*, PNNL-18427, Pacific Northwest National Laboratory, Richland, Washington, 2009.
49. J. P. McKinley, C. J. Zeissler, J. M. Zachara, R. J. Serne, R. M. Lindstrom, H. T. Schaef and R. D. Orr, *Environ. Sci. Technol.*, 2001, **35**, 17.

50. J. M. Zachara, S. C. Smith, C. X. Liu, J. P. McKinley, R. J. Serne and P. L. Gassman, *Geochim. Cosmochim. Acta*, 2002, **66**, 2.
51. R. J. Serne, G. V. Last, H. T. Schaef, D. C. Lanigan, C. W. Lindenmeier, C. C. Ainsworth, R. E. Clayton, V. L. LeGore, M. J. O'Hara, C. F. Brown, R. D. Orr, I. V. Kutnyakov, T. C. Wilson, K. B. Wagnon, B. A. Williams and D. S. Burke, *Characterization of Vadose Zone Sediment: Borehole 41-09-39 in the S-SX Waste Management Area*, PNLL-13757-3, Pacific Northwest National Laboratory, Richland, Washington, 2002.
52. POST, *POST, Parliamentary Office of Science and Technology, Depleted Uranium Postnote number 154*, 2001, www.parliament.uk/documents/post/pn154.pdf (accessed 8th August 2010).
53. M. Fan, T. Thongsri, L. Axe and T.A. Tyson, *Chemosphere*, 2005, **60**, 1.
54. W. Dong, G. Xie, T. R. Miller, M. P. Franklin, T. P. Oxenberg, E. J. Bouwer, W. P. Ball and R. U. Halden, *Environ. Pollut.*, 2006, **142**, 1.
55. UNEP, *Depleted Uranium in Kosovo: Postconflict environmental assessment*, United Nations Environment Programme, Imprimerie Chirat, France, 2001.
56. P. Diehl, *WISE uranium project, fact sheet.*, 2001.
57. The Royal Society, *The Royal Society, The health hazards of depleted uranium munitions - Part I Policy document 6/01*, The Royal Society, London, The Royal Society 2001.
58. S. Handley-Sidhu, M. J. Keith-Roach, J. R. Lloyd, and D. J. Vaughan, *Sci. Total Environ.*, In Press, Corrected Proof. .
59. J. R. Lloyd, *FEMS Microbiol. Rev.*, 2003, **27**, 2–3.
60. Y. A. Gorby, F. Caccavo and H. Bolton, *Environ. Sci. Technol.*, 1998, **32**, 2.
61. D. R. Lovley, E. J. P. Phillips, Y. A. Gorby and E. R. Landa, *Nature*, 1991, **350**, 6317.
62. D. R. Lovley and E. J. Phillips, *Appl. Environ. Microbiol.*, 1988, **54**, 6.
63. J. R. Lloyd and D. R. Lovley, *Curr. Opin. Biotechnol.*, 2001, **12**, 3.
64. W. P. de Bruin, M. J. Kotterman, M. A. Posthumus, G. Schraa and A. J. Zehnder, *Appl. Environ. Microbiol.*, 1992, **58**, 6.
65. G. C. Barbee, *Ground Water Monit. R.*, 1994, **14**, 1.
66. R. Oldenhuis, R. L. Vink, D. B. Janssen and B. Witholt, *Appl. Environ. Microbiol.*, 1989, **55**, 11.
67. P. Pant and S. Pant, *J. Environ. Sci. - China*, 2010, **22**, 1.
68. C. White and G. M. Gadd, *Microbiol. - UK*, 1996, **142**.
69. M. Ledin and K. Pedersen, *Earth-Sci. Rev.*, 1996, **41**, 1–2.
70. C. White, A. K. Sharman and G. M. Gadd, *Nat. Biotechnol.*, 1998, **16**, 6.
71. M. J. Beazley, R. J. Martinez, P. A. Sobecky, S. M. Webb and M. Taillefert, *Environ. Sci. Technol.*, 2007, **41**, 16.
72. L. E. Macaskie, R. M. Empson, A. K. Cheetham, C. P. Grey and A. J. Skarnulis, *Science*, 1992, **257**, 5071.
73. G. M. Gadd, *Geoderma*, 2004, **122**, 2–4.
74. J. C. Renshaw, J. R. Lloyd and F. R. Livens, *C. R. Chim.*, 2007, **10**, 10–11.
75. G. M. Gadd, *Curr. Opin. Biotechnol.*, 2000, **11**, 3.
76. J. R. Lloyd, R. T. Anderson, and L. E. Macaskie, in "Bioremediation: Applied Microbial Solutions for Real-World Environmental Cleanup",

(Eds.) R. M. Atlas and J. Philip, ASM Press, Washington D.C., USA. 2005, 293.
77. A. Mrozik and Z. Piotrowska-Seget, *Microbiol. Res.*, **165**, 5.
78. D. Stephenson and T. Stephenson, *Biotechnol. Adv.*, 1992, **10**, 4.
79. T. F. Guerin, *J. Hazard. Mater.*, 1999, **65**, 3.
80. T. F. Guerin, *J. Hazard. Mater.*, 2008, **154**, 1–3.
81. S. B. Larsen, D. Karakashev, I. Angelidaki and J. E. Schmidt, *J. Hazard. Mater.*, 2009, **164**, 2–3.
82. K. E. Gerhardt, X.-D. Huang, B. R. Glick and B. M. Greenberg, *Plant Sci.*, 2009, **176**, 1.
83. I. Alkorta and C. Garbisu, *Bioresour. Technol.*, 2001, **79**, 3.
84. R. L. Siegrist, M. A. Urynowicz and O. R. West, *Ground Water Currents - Developments in Innovative Ground Water Treatment*, 2000, **37**, 1–3.
85. K. M. Jerome, B. Riha and B. B. Looney, *Final Report for Demonstration of In Situ Oxidation of DNAPL Using the Geo-Cleanse® Technology*, U.S. Department of Energy, Westinghouse Savannah River Company, Aiken, South Carolina, 1997.
86. A. B. Cundy, L. Hopkinson and R. L. D. Whitby, *Sci. Total Environ.*, 2008, **400**, 1–3.
87. R. Thiruvenkatachari, S. Vigneswaran and R. Naidu, *J. Ind. Eng. Chem.*, 2008, **14**, 2.
88. Y. Furukawa, J.-w. Kim, J. Watkins and R. T. Wilkin, *Environ. Sci. Technol.*, 2002, **36**, 24.
89. W. R. Richmond, M. Loan, J. Morton and G. M. Parkinson, *Environ. Sci. Technol.*, 2004, **38**, 8.
90. D. W. Blowes, R. W. Gillham, C. J. Ptacek, R. W. Puls, T. A. Bennett, S. F. O'Hannesin, C. J. Hanton-Fong and J. G. Bain, *An In Situ Permeable Reactive Barrier for the Treatment of Hexavalent Chromium and Trichloroethylene in Ground Water: Volume 1 Design and Installation*, U.S. Environmental Protection Agency, Washington D.C., DC, USA, EPA/600/R-99/095a, 1999.
91. R. M. Powell, D. W. Blowes, R. W. Gillham, D. Schultz, T. Sivavec, R. W. Puls, J. L. Vogan, P. D. Powell and R. Landis, *Permeable reactive Barrier Technologies for Contaminant Remediation*, U.S. Environmental Protection Agency, Washington D.C., DC, USA, EPA/600/R-98/125, 1998.
92. A. D. Henderson and A. H. Demond, *Environ Eng Sci*, 2007, **24**, 4.
93. C. N. Mulligan, R. N. Yong and B. F. Gibbs, *J. Hazard. Mater.*, 2001, **85**, 1–2.
94. C. Löser, A. Zehnsdorf, P. Hoffmann and H. Seidel, *Chemosphere*, 2007, **66**, 9.
95. R. W. Peters, *J. Hazard. Mater.*, 1999, **66**, 1–2.
96. A. Polettini, R. Pomi, E. Rolle, D. Ceremigna, L. De Propris, M. Gabellini and A. Tornato, *J. Hazard. Mater.*, 2006, **137**, 3.
97. G. Dermont, M. Bergeron, G. Mercier and M. Richer-Laflèche, *J. Hazard. Mater.*, 2008, **152**, 1.
98. W. H. Zhang, H. Huang, F. F. Tan, H. Wang and R. L. Qiu, *J. Hazard. Mater.*, 2010, **173**, 1–3.

99. C. N. Mulligan, R. N. Yong and B. F. Gibbs, *Eng. Geol.*, 2001, **60**, 1–4.
100. J. Virkutyte, M. Sillanpää and P. Latostenmaa, *Sci. Total Environ.*, 2002, **289**, 1–3.
101. R. Lageman, R. L. Clarke and W. Pool, *Eng. Geol.*, 2005, **77**, 3–4.
102. A. B. Cundy and L. Hopkinson, *Appl. Geochem.*, 2005, **20**, 5.
103. K. B. Gregory and D. R. Lovley, *Environ. Sci. Technol.*, 2005, **39**, 22.
104. J. C. Thrash and J. D. Coates, *Environ. Sci. Technol.*, 2008, **42**, 11.
105. V. Vermeul, B. Bjornstad, B. Fritz, J. Fruchter, R. Mackley, D. Mendoza, D. Newcomer, M. Rockhold, D. Wellman and M. Williams, *300 Area Uranium Stabilization Through Polyphosphate Injection: Final Report*, PNNL–18529, Pacific Northwest National Laboratory, Richland, Washington, WA, USA, 2009.
106. D. Wellman, E. Pierce, D. Bacon, M. Oostrom, K. Gunderson, S. Webb, C. Bovaird, E. Cordova, E. Clayton, K. Parker, R. Ermi, S. Baum, V. Vermeul and J. Fruchter, *300 Area Treatability Test: Laboratory Development of Polyphosphate Remediation Technology for In Situ Treatment of Uranium Contamination in the Vadose Zone and Capillary Fringe*, PNNL-17818, Pacific Northwest National Laboratory, Richland, Washington, 2008.
107. D. Wellman, E. Pierce, M. Oostrom and J. Fruchter, *Experimental Plan: 300 Area Treatability Test: In Situ Treatment of the Vadose Zone and Smear Zone Uranium Contamination by Polyphosphate Infiltration*, PNNL-16823, Pacific Northwest National Laboratory, Richland, Washington, WA, USA, 2007.
108. J. Jeanjean, J. Rouchaud, L. Tran and M. Fedoroff, *J. Radioanal. Nucl. Chem.*, 1995, **201**, 6.
109. C. Y. Shen and F. W. Morgan, in *Environmental Phosphorous Handbook*, E. J. Griffith, *et al.*, John Wiley & Sons, New York, 1973.
110. K. T. Finneran, R. T. Anderson, K. P. Nevin and D. R. Lovley, *Soil Sediment Contam.*, 2002, **11**, 3.
111. H. A. Vrionis, R. T. Anderson, I. Ortiz-Bernad, K. R. O'Neill, C. T. Resch, A. D. Peacock, R. Dayvault, D. C. White, P. E. Long and D. R. Lovley, *Appl. Environ. Microbiol.*, 2005, **71**, 10.
112. A. L. N'Guessan, H. A. Vrionis, C. T. Resch, P. E. Long and D. R. Lovley, *Environ. Sci. Technol.*, 2008, **42**, 8.
113. N. E. Korte, *Zero-Valent Iron Permeable reactive Barriers: A Review of Performance*, Publication No. 5056, Environmental Sciences Division Oak Ridge National Laboratory, 2001.
114. D. B. Watson, D. H. Phillips and B. Gu, *Performance Evaluation of in-situ Iron Reactive Barriers at the Oak Ridge Y-12 Site*, ORNL/TM-2001/193, Oak Ridge National Laboratory, 2002.
115. W. M. Wu, J. Carley, T. Gentry, M. A. Ginder-Vogel, M. Fienen, T. Mehlhorn, H. Yan, S. Caroll, M. N. Pace, J. Nyman, J. Luo, M. E. Gentile, M. W. Fields, R. F. Hickey, B. H. Gu, D. Watson, O. A. Cirpka, J. Z. Zhou, S. Fendorf, P. K. Kitanidis, P. M. Jardine and C. S. Criddle, *Environ. Sci. Technol.*, 2006, **40**, 12.

Decommissioning of Nuclear Sites

ANTHONY W. BANFORD* AND RICHARD B. JARVIS

ABSTRACT

This paper discusses the various phases of decommissioning and what they seek to achieve, and discusses some drivers for decommissioning both immediately after operations cease and after a period of time. The scale of the decommissioning challenge in the UK is then outlined and some criteria for the selection of the best decommissioning option for a facility are introduced. Finally, the potential environmental impacts of each of the different decommissioning stages are discussed.

1 Introduction

Nuclear decommissioning can be defined as "The process whereby a nuclear facility, at the end of its economic life, is taken permanently out of service and its site made available for other purposes".[1] The process of decommissioning incurs financial costs known as liabilities, both from the decommissioning process itself and the associated waste management and environmental remediation. The United Kingdom has played a major role in the development of nuclear power and has a large number of facilities which will ultimately need to be decommissioned.

2 The Goal of Decommissioning

Decommissioning aims to take a plant which has been washed out at the end of operations and leave the site in its planned end state. End states could range from buildings containing waste entombed in concrete in their current location

*Corresponding author

Issues in Environmental Science and Technology, 32
Nuclear Power and the Environment
Edited by R.E. Hester and R.M. Harrison

Published by the Royal Society of Chemistry, www.rsc.org

to completely clear sites available for another use. Intermediate states might include building foundations left in place while above ground structures are removed from the site.

Materials which are exported from the site must be treated to make them acceptable for final disposal. The UK presently has a near-surface, low level waste repository (LLWR) and is planning to build a Geological Disposal Facility (GDF) to contain intermediate and high level wastes. In some circumstances materials will be sufficiently inactive that they are suitable for free-release and do not require disposal to one of these facilities. Disposal to one of the repositories typically requires wastes to be boxed and filled with grout, or other encapsulant. Waste boxes are stacked in the repository and then further material will be introduced to surround the boxes and provide a controlled local chemical environment. One goal of decommissioning is to minimise the amount of waste going to the GDF and LLWR.

The drivers for decommissioning are both varied and interrelated. Decommissioning can therefore proceed for a number of reasons. These include:

- Reduction of the hazard associated with legacy radioactive inventory in ageing facilities.[2]
- Completion of decommissioning allows sale or reuse of the site.
- Continued existence of the site requires conformance to safety legislation and so requires care and maintenance that will incur ongoing costs that could be avoided by completion of decommissioning.
- There is a duty to future generations to avoid leaving them to clean up the waste generated by the current generation.
- To remove visual intrusion caused by large facilities.

Given that a facility is to be decommissioned, there is then a decision as to when this shall take place. The factors above will strongly influence the prioritisation of decommissioning but a number of other factors should also be considered:

- Prompt decommissioning means that site knowledge is retained and can be used to assist in the decommissioning process.
- Availability of money to perform decommissioning may be limited.
- Financial depreciation makes it attractive to defer large expenditure.
- Radioactive decay means that decommissioning may be easier, and therefore cheaper, if it is deferred. For some wastes, the radioactive decay may result in material being reclassified as a less onerous waste form, so requiring less treatment and incurring reduced costs.
- A final disposal route for wastes may not be available. This may require construction of intermediate facilities incurring costs and using natural resources that would not otherwise be required; these facilities themselves will require decommissioning at the end of their lives. It may therefore be desirable to postpone decommissioning until a final disposal route is available.

- Enhanced technology may be developed during the deferral period which improves the process.

3 Stages of Decommissioning

In the United Kingdom, the Nuclear Decommissioning Authority (NDA) is a non-departmental government body founded in 2005 to manage the UK civil nuclear wastes. The NDA defined a number of stages of the decommissioning process as illustrated in Figure 1.[3]

Whilst each plant and site may have their own characteristics, this series of generic stages provide a good overview of decommissioning activities from the end of operations through to site closure.

The first stage after normal plant operations cease is known as Post-Operational Clean Out (POCO). In the case of a reactor site this is removal of the fuel from the reactor; for other facilities it typically requires existing equipment, with only minor modifications, to be used by plant operators to move most of the radioactivity out of the plant. POCO will typically use only chemicals (and equipment) that were used during plant operation, and utilise existing waste and effluent treatment routes.

The next stage, Initial Decommissioning, removes or fixes loose radioactive material within pipework and vessels to reduce dose rates and ease access to facilitate further decommissioning tasks. This may use special cleaning chemicals and so require additional effluent treatment equipment. The transition from POCO to initial decommissioning may involve changes in staff and controlling procedures and so is potentially problematic; particular advice is available for making this transition.[4]

The Surveillance and Maintenance stage applies only to facilities that are not in a passively safe state following Initial Decommissioning and which require a period of Surveillance and Maintenance prior to Interim Decommissioning. In these cases certain plant systems would remain operational (*e.g.* services,

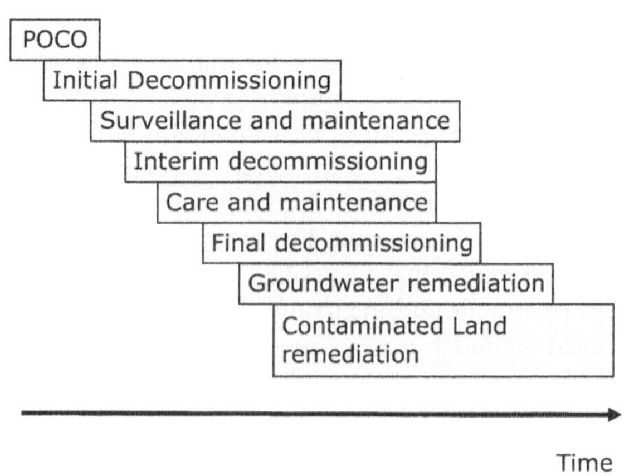

Figure 1 The stages of decommissioning.

radiological monitoring and ventilation systems), maintenance regimes would remain in place and some plant enhancement may be necessary to maintain building structural integrity.

The Interim Decommissioning stage is when the work required to convert a facility to a passively safe state is carried out. Typically this would involve removal of residual radioactive inventory from the plant, dismantling and removal of plant and equipment, removal of non-radioactive facilities and, where possible, reduction of the building footprint. At the end of this stage, the plant will be in a passively safe state with systems and processes de-energised, deactivated and drained.

The Care and Maintenance stage allows limited monitoring and observation of a facility prior to final decommissioning. It can be distinguished from surveillance and maintenance because few resources are required. Typically care and maintenance might be used to allow levels of radioactivity to decay, but it is also possible that the facility has been made sufficiently safe that resources are most effectively used on other facilities at a given time. Effort to maintain the plant in this state would be minimal, confined to routine monitoring and surveillance of the facility and the building fabric with very few, if any, operators dedicated to the plant on a full-time basis. It is important the facility is adequately enclosed during this period and some guidance for this is provided.[5]

Final Decommissioning will bring a plant or facility to its agreed end-point, including final site clearance but excluding any contaminated land or groundwater remediation. This includes final dismantling of installed plant and equipment, strip-out of any remaining facilities within the building and demolition of cells, internal structures and the building envelope. All wastes generated will be disposed of or stored awaiting disposal. The end-point reached at the completion of this phase will be such that any danger or hazard that may remain to workers, the general public or the environment is at a minimum level consistent with the principles of ALARP (As Low As Reasonably Practicable).

Subsequent work would consider both Groundwater Remediation and Contaminated Land Remediation prior to site close-out.

4 The Scale of the Decommissioning Challenge in the UK

There are a number of nuclear licensed sites in the UK. The NDA owns 20 sites spread across the UK. These sites were formally operated by BNFL and UKAEA, and date back to the start of the UK nuclear programme. There are also further sites operated by other organisations, notably British Energy which operates the Advanced Gas Cooled Reactor (AGR) sites and the Pressurised Water Reactor (PWR) at Sizewell.

The NDA sites contain a wide range of facilities and decommissioning challenges. The sites comprise:

- Fuel recycle plants at Sellafield
- Fuel enrichment plant at Capenhurst

- Twelve reactor sites (five in each of Magnox North and Magnox South, two included within Sellafield)
- A research reactor site at Dounreay
- Two research sites (Harwell and Winfrith)
- A fuel manufacture plant at Springfields
- The Low Level Waste Repository (LLWR).

These facilities will produce the majority of the waste that must be managed in the UK but contributions will also arise from British Energy reactors and others.

The total liabilities associated with these NDA sites are shown in Figure 2.[6] Direct decommissioning activities total around £10bn while associated waste management activities make up a proportion of a further £10bn. The total discounted liability for the UK is around £45bn; internationally the challenge is even greater.[7]

The sites vary markedly in the scale and nature of the decommissioning required. Sellafield is responsible for handling highly radioactive spent fuel and has a correspondingly high decommissioning liability. Some of the plants at Sellafield date back to the early years of the UK nuclear industry and decommissioning these old structures in a safe manner is challenging. Reactor sites, once defueled, have a greatly reduced radioactive burden, which can be further reduced by allowing a "safe store" period of up to 75 years. Reactors therefore form a smaller contribution to the decommissioning liability than Sellafield. Springfields manufactures fuel and does not handle spent, highly active fuel, and so poses less of a decommissioning challenge and requires

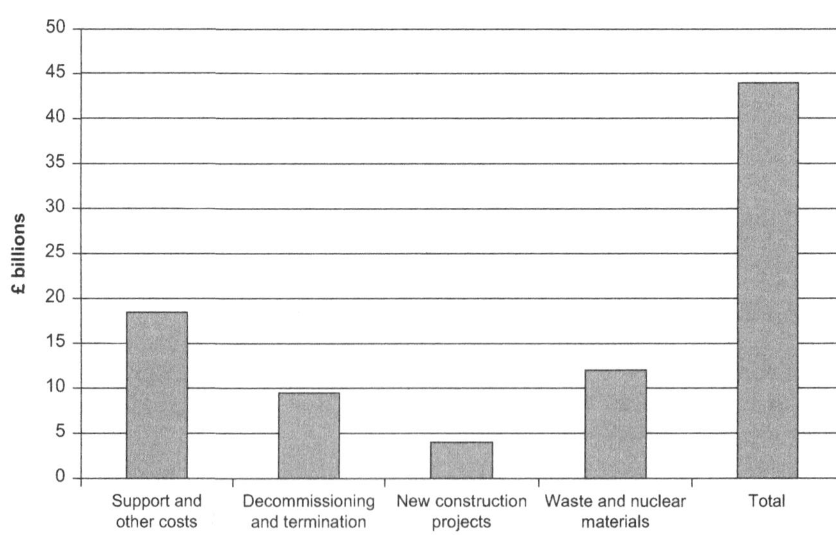

Figure 2 Total discounted nuclear liability (discounted at 2.2%).

Table 1 CoRWM inventory.

Type	*Packaged Volume (m^3)*	*Radioactivity (TBq)*
HLW	1290	39 000 000
ILW	353 000	2 400 000
Plutonium	3270	4 000 000
Uranium	74 950	3000
Spent nuclear fuel	8150	33 000 000
Total	477 860	78 000 000

correspondingly less money to complete. Research sites pose challenges not present on the other sites, for example Dounreay has stored large quantities of liquid sodium alloy used as a coolant in a test reactor which now must be treated.

The committee on radioactive waste management (CoRWM) has produced an inventory of materials requiring management.[8] Their findings are shown in Table 1.

The high level waste (HLW) and recovered uranium and plutonium derive from operations and POCO at Sellafield and so are not associated with decommissioning. The key wastes arising from decommissioning will be intermediate level waste (ILW), low level waste (LLW) and very low level waste (VLLW).

The Sellafield integrated waste strategy[9] suggests that around half of the ILW arising at Sellafield will be associated with decommissioning; this is around 140 000 m^3, while the Magnox South integrated waste strategy[10] suggests about 26 000 m^3 of ILW will arise from decommissioning. Further contributions will be made from Dounreay 9000 m^3[11] and from Sellafield contaminated land 1600 m^3.[9] In total at least 60% of the ILW requiring management derives from decommissioning – over 200 000 m^3.

Much of the waste arising from decommissioning will be low level waste (LLW) or very low level waste (VLLW). Predicted volume arisings beyond 2030 are dominated by decommissioning activities.[12] Total volume arisings are shown in Figure 3 while Figure 4 shows the breakdown into different materials. It is clear from Figure 4 that much of the VLLW material is soil and rubble from decommissioning operations, it would also be expected that a significant proportion of the metals would be associated with vessels and metal reinforcement associated with decommissioning. Overall perhaps around 75% of the VLLW (around 1.3 million m^3) results from decommissioning operations. A smaller, but still significant proportion of the LLW is also associated with decommissioning.

5 Decommissioning Techniques

One of the key decisions for decommissioning is whether it will be performed remotely or will be hands-on, with operators using tools directly. It will often be cheaper and easier to use manual techniques since this allows maximum

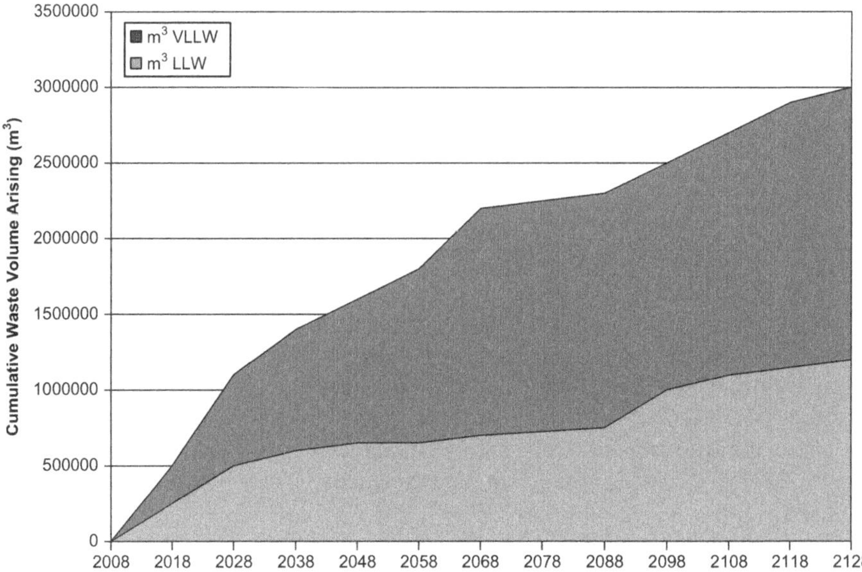

Figure 3 Raw cumulative arisings of LLW and VLLW.

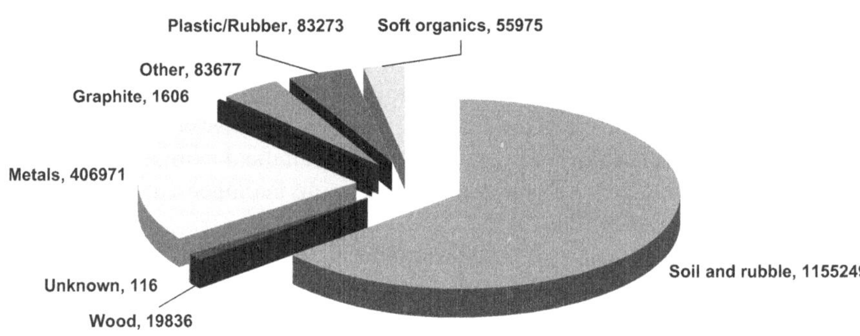

Figure 4 Material in raw VLLW (m^3).

flexibility in choosing appropriate tools and allows parallel working to reduce the time required for decommissioning. However, manual decommissioning may not be possible if radiation levels are too high.

Manual decommissioning can exploit hand tools as well as sit-on machinery. Remote decommissioning uses manipulators and other tools mounted on cranes or remotely-operated-vehicles and directed by operators using a camera for guidance.

Decommissioning may try to separate different waste categories such as intermediate and low level wastes from free release material. In some cases physical techniques such as cutting, scabbling a surface, water jetting or sand blasting might be used.[13] In other instances chemical decontamination might offer advantages. In each case the cost and environmental benefits of performing the operation must be weighed against the cost, resource usage and dose incurred by further processing the waste. Some methods for decontamination are discussed.[14] Decontamination to support decommissioning can use more aggressive chemicals than are used for cleaning equipment prior to maintenance during plant operations, it is also possible to deploy decontamination either *in situ* prior to removal of items or *ex situ* after they are removed from their original location.

Examples of the successful decontamination of streams are the Berkeley fuelling machines: 1700 tonnes of which were recycled, 60 tonnes of which were disposed as LLW, while 30 tonnes remain in store awaiting disposal. Chemical cleaning of the Berkeley gas ducts allowed 750 tonnes of steel to be recycled.

6 Selection of a Decommissioning Approach

There are a number of decisions to be made for each building that is to be decommissioned. This section outlines a set of criteria that can be used to select a preferred decommissioning approach.[15] These criteria reflect the issues discussed earlier. A range of decommissioning approaches that may be assessed by this approach are discussed[16–18] and include issues of prompt decommissioning *versus* deferred decommissioning as well as selection of an end state.

The criteria are shown in Figure 5. The criteria form a hierarchy which, at the top level, has the three pillars of sustainability. These objectives are expanded into criteria and sub-criteria. Environmental impact is divided into radiological impact on man and the environment, resource usage, non-radiological discharges, local intrusion (which includes such factors as noise and visual pollution) and hazard potential (a measure developed as part of the NDA prioritisation process).[2,19]

These criteria provide a complete set of issues for the assessment of decommissioning projects, however, for any given assessment it may be useful to omit criteria which are not significantly different between all options under consideration, and it may also be convenient to subdivide other criteria to make best use of more readily available metrics.

It is important that the criteria are assessed over the whole lifecycle of each proposed decommissioning approach and that the end points are the same in each case in order to obtain the preferred solution.

If this approach is followed then optimum approaches to decommissioning will be obtained. Considering a complex site with many facilities or the UK as a whole, deferred decommissioning may be required in order to make the programme of work fit within the available annual budget. Such issues can be investigated using the same criteria and the impact in terms of cost, environment and social factors assessed for leaving a plant in a state of surveillance and maintenance or care and maintenance.

By these methods programmes of work delivering maximum environmental benefit as quickly as possible can be derived.

Objective 1: Environment and Safety

Criterion 1: Environment and Public Safety

Sub-criterion 1.1: Radiological Impact - Man

Sub-criterion 1.2: Radiological Impact - Environment

Sub-criterion 1.3: Resource Usage

Sub-criterion 1.4: Non-radiological Discharges

Sub-criterion 1.5: Local Intrusion

Sub-criterion 1.6: Hazard Potential

Criterion 2: Worker Safety

Sub Criterion 2.1: Radiological Worker Safety

Sub Criterion 2.2: Non-radiological Worker Safety

Criterion 3: Security

Objective 2: Economic

Criterion 4: Economic Cost and Benefit

Sub-criterion 4.1: Cost

Sub-criterion 4.2: Spin-off

Criterion 5: Technology Predictability

Sub-criterion 5.1: Concept Predictability

Sub-criterion 5.2: Operational Predictability

Objective 3: Social

Criterion 6: Stability of Employment

Criterion 7: Burden on Future Generations

Figure 5 Criteria for the selection of a decommissioning approach.

7 Environmental Impacts of Decommissioning

The different stages of decommissioning were presented in section 3. This section summarises the potential environmental effects of performing decommissioning (see Table 2) and also potential environmental impacts posed by the facilities between the different stages of decommissioning (see Table 3).

Each of the decommissioning stages may result in discharges to the environment. For example demolition work will produce dusts and decontamination

Table 2 Environmental impacts of decommissioning.

	Discharges	*Use of resources*	*HLW/ILW waste disposal*	*LLW/VLLW waste disposal*
POCO	Significant discharges	Use existing effluent treatment plants and chemicals	Significant arisings to existing waste routes	Low
Initial Decommissioning	Some discharges	May require new effluent treatment plant; may require aggressive cleaning chemicals	Some arisings to existing/new waste routes	Low
Surveillance and Maintenance	Ongoing discharges to service streams	Ongoing provision of services and associated clean up equipment	Very low arisings of secondary waste	Low arisings of secondary waste
Interim Decommissioning	Some discharges	Equipment for vessel/pipework removal and size reduction required	Primary wastes generated: dismantled vessels and pipework. Wastes require packaging.	Primary wastes generated: dismantled vessels and pipework. Wastes require packaging.
Care and Maintenance	Negligible	Negligible	Negligible	Negligible
Final Decommissioning	Low activity dusts and cleaned liquors	Significant use of retrievals equipment and waste packaging	Small quantities expected under some facilities	Very large quantities expected
Groundwater Remediation and Contaminated Land	Low	Significant use of retrievals equipment and waste packaging	Small quantities expected under some facilities	Very large quantities expected

Table 3 Environmental impacts of buildings between decommissioning stages.

Potential for release in maloperation	*Requirement for services with associated discharges and secondary waste*	*Potential migration of species from contaminated land*
Following completion of operations		
Large quantities of mobile species may be released in maloperation conditions	Water, steam, electricity and process air may all bc required to maintain a safe building	Potential migration of species from contaminated land
POCO		
Mobile species may be released in maloperation conditions; quantity greatly reduced by POCO	Water, steam, electricity and process air may all be required to maintain a safe building	Potential migration of species from contaminated land
Initial Decommissioning		
Greatly reduced risk of release; structural damage required to release activity	Services may be reduced but not eliminated	Potential migration of species from contaminated land
Surveillance and Maintenance		
Further reduced impact of release due to radioactive decay	Services may be reduced but not eliminated	Potential migration of species from contaminated land; impact of radioactive species reduced by decay
Interim Decommissioning		
Further reduced impact of release due to retrieval of residuals; major structural damage required to release activity	Services no longer required	Potential migration of species from contaminated land
Care and Maintenance		
Further reduced impact of release due to radioactive decay; major structural damage required to release activity	Services no longer required	Potential migration of species from contaminated land; impact of radioactive species reduced by radioactive decay
Final Decommissioning		
Building removed; no further risk	Services no longer required	Potential migration of species from contaminated land
Ground water remediation and Contaminated Land Remediation		
Building removed; no further risk	Services no longer required	Water and land remediated; no further risk

may produce liquid effluents. In each case suppression of the effluent production may result in formation of secondary wastes, such as filters, which then require ultimate disposal. Natural resources are also required for waste packaging and construction of decommissioning equipment, some of which may not be able to be reused.

In between decommissioning stages, environmental impacts include the need for provision of services such as water and air and the treatment of the associated discharges; clean up and secondary waste generation; potential discharges in the event of a major building failure, perhaps as a result of a natural disaster or malicious action; and migration of species in the ground under the building from previous spillages.

8 Conclusions

Decommissioning is an important phase in the lifecyle of any nuclear facility, covering the transition from an operating facility to its planned end state. The United Kingdom has had a significant civil nuclear operation for many years, and as such has a significant decommissioning challenge associated with both fuel cycle plant and reactors. The decommissioning of legacy plant involves significant financial liabilities and involves large volumes of waste.

Whilst the process of decommissioning a facility can be described generically as a series of stages, the selection of a decommissioning strategy is typically plant, site or region specific. Decommissioning can be driven by many factors, ranging from a desire to reduce the hazard associated with an ageing facility through to a need to release the site for re-use. This paper has highlighted a series of criteria that should be considered when selecting a decommissioning strategy, of which environment and safety factors are of fundamental importance.

References

1. DTI, 2002. *Managing the Nuclear Legacy: a strategy for action*. London: Department of Trade and Industry, Cm 5552.
2. NDA, 2008. *NDA prioritisation procedure*. Moor Row, Cumbria: Nuclear decommissioning authority, EGPR02, Rev 4.
3. NDA, 2010. *Project controls framework document*. Moor Row, Cumbria: Nuclear decommissioning authority, PCP-M, Rev 1.
4. IAEA, 2004. *Safety considerations in the transition from operation to decommissioning of nuclear facilities*. Vienna: International Atomic Energy Agency, Safety report series 36.
5. IAEA, 2002. *Safe enclosure of nuclear facilities during deferred dismantling*. Vienna: International Atomic Energy Agency, Safety reports series 26.
6. NDA, 2009. *Nuclear decommissioning authority annual report and accounts 2008/2009*. Moor Row, Cumbria: Nuclear decommissioning authority.
7. IAEA, 2004. *Status of the decommissioning of nuclear facilities around the world*. Vienna, International Atomic Energy Agency, STI/PUB/1201.

8. CoRWM, 2006. *Managing our radioactive waste safely: CoRWM's recommendations to government.* London: Committee on Radioactive Waste Management, CoRWM Doc 700.
9. Sellafield Ltd, 2008. *Sellafield integrated waste strategy – update to version 2.* Seascale, Cumbria: Sellafield Ltd.
10. Magnox South, 2008. *Magnox south integrated waste strategy.* Berkeley, Gloucestershire: Magnox South Sites, LTP08 Rev D, MES/EST/GEN/REP/0002/09.
11. UKAEA, 2008. *Dounreay 'interim' integrated waste strategy.* Dounreay: UKAEA, WSU/STRATEGY/P033(08), Issue 4.
12. LLW Repository Ltd, 2009. *Strategic BPEO study for very low level waste.* Holmrook, Cumbria: LLW Repository Ltd.
13. IAEA, 2008. *Innovative and adaptive technologies in decommissioning of nuclear facilities.* Vienna: International Atomic Energy Agency, IAEA TECHDOC 1602.
14. IAEA, 1999. *State of the art technology for decontamination and dismantling of nuclear facilities.* Vienna: International Atomic Energy Agency, Technical reports series 395.
15. Banford, A. W., Eccles, H., Jarvis, R. B. and Ross, D. N., 2010. *The development of an integrated waste management approach for irradiated graphite.* In Proceedings of Waste Management 2010, March 7–11th 2010, Phoenix, AZ. Tempe, AZ: WM Symposia, 3975–3987.
16. IAEA, 2007. *Decommissioning strategies for facilities using radioactive material.* Vienna: International Atomic Energy Agency, Safety report series 50.
17. IAEA, 2001. *Decommissioning of nuclear fuel cycle facilities.* Vienna, International Atomic Energy Agency, WS-G-2.4.
18. IAEA, 1999. *Decommissioning of nuclear power plants and research reactors.* Vienna, International Atomic Energy Agency, WS-G-2.1.
19. Jarvis, R., 2006. *The Radiological Hazard Potential: a progress measure for nuclear cleanup.* Moor Row, Cumbria, NDA, EGR009.

Geodisposal of Higher Activity Wastes

KATHERINE MORRIS*, GARETH T. W. LAW AND NICK D. BRYAN

ABSTRACT

In the UK, there is a nuclear waste legacy associated with over 50 years of nuclear power generation that is currently stored at the Earth's surface. This is a global phenomenon in which many nations are now facing up to the radioactive waste legacy of several decades of nuclear power generation. As society considers new nuclear power as a low carbon, secure source of energy, it is apparent that geological disposal of higher activity radioactive wastes is now the favoured route for management of this highly radioactive legacy material. Timely implementation of geological disposal is therefore a current challenge facing the UK and other nuclear nations if we are to demonstrate safe management of these materials for future generations. In this chapter, we review the type and characteristics of the higher activity wastes that the UK needs to dispose of; examine the concept of a geological disposal facility in the context of UK and international experience; and discuss the proposed implementation pathway for UK higher activity waste geodisposal in the context of our large and complex nuclear legacy. Finally, we discuss the environmental chemistry research challenges that we see as vital to the safe management and disposal of these legacy radioactive wastes.

1 Introduction

Radioactive waste management is now a pressing issue for the UK: there is an extensive legacy of higher activity wastes (HAW), some of which have been treated for storage and geological disposal; there is also a need to demonstrate

*Corresponding author

Issues in Environmental Science and Technology, 32
Nuclear Power and the Environment
Edited by R.E. Hester and R.M. Harrison

Published by the Royal Society of Chemistry, www.rsc.org

that wastes generated from any new nuclear power reactors which are proposed as a secure, low-carbon energy source, can be managed in the future. In England and Wales, geological disposal has been chosen as the long-term management pathway for HAWs. The implementation programme for a national geological disposal facility (GDF) was launched in 2008 (ref. 1). Indeed, the scientific and societal challenges facing the UK in legacy management are being echoed at a global scale. In this chapter, we discuss the types and quantities of radioactive wastes that are likely to be disposed of to a GDF and the likely planning and design concept for this type of facility. We also highlight selected environmental chemistry research challenges in geological disposal as they relate to the reduction in uncertainties for the GDF safety case.

2 Radioactive Wastes

Radioactive wastes comprise materials that are contaminated by, or incorporate, radioactivity above threshold levels defined in UK legislation, and for which there is no further economic use. It is important to note that the UK radioactive waste legacy is complex and varied because of the diverse history of power reactors and nuclear weapons development. The HAWs destined for geodisposal in the UK are identified as items that: (i) cannot be managed under the policy for the long term management and storage of solid low level radioactive waste in the United Kingdom, and (ii) are not managed under the Scottish Executive's emerging policy for radioactive waste.[1] These HAWs are broadly categorised depending on their radioactive content as either heat-generating high level waste (HLW), or non heat-generating intermediate level waste (ILW). In addition to HLW and ILW, some small volume items of low level waste (LLW), a range of materials that are currently not declared as waste such as spent fuels and stockpiles of uranium and plutonium, and any wastes from any new nuclear reactors, may be managed *via* geodisposal. Each waste type is described in detail in sections 2.1–2.4.

2.1 High Level Wastes

In the UK, HLWs are defined as materials that are heat generating due to their radioactivity, and this heat generation needs to be considered during the design of storage and disposal facilities.[2] HLW in the UK is dominated by materials derived from spent nuclear fuel reprocessing (see Figure 1).[2] Prior to reprocessing, spent fuel is stored for several years in a holding pond to allow short lived radionuclides to undergo radioactive decay and to dissipate the resultant heat. At the Sellafield site, spent fuel from a range of reactor types is either taken out of its cladding or chopped up, dissolved in nitric acid and is chemically processed in the PUREX process, to separate uranium, plutonium and waste fission products. During reprocessing, the fission products are ultimately separated into a solution phase which is known as highly active raffinate (HAR). This fission product laden nitric acid solution contains the bulk of the radioactivity that was associated with the spent fuel. It will continue to generate heat for decades and is dominated by radioisotopes such as ^{90}Sr and ^{137}Cs. The HAR is converted to a

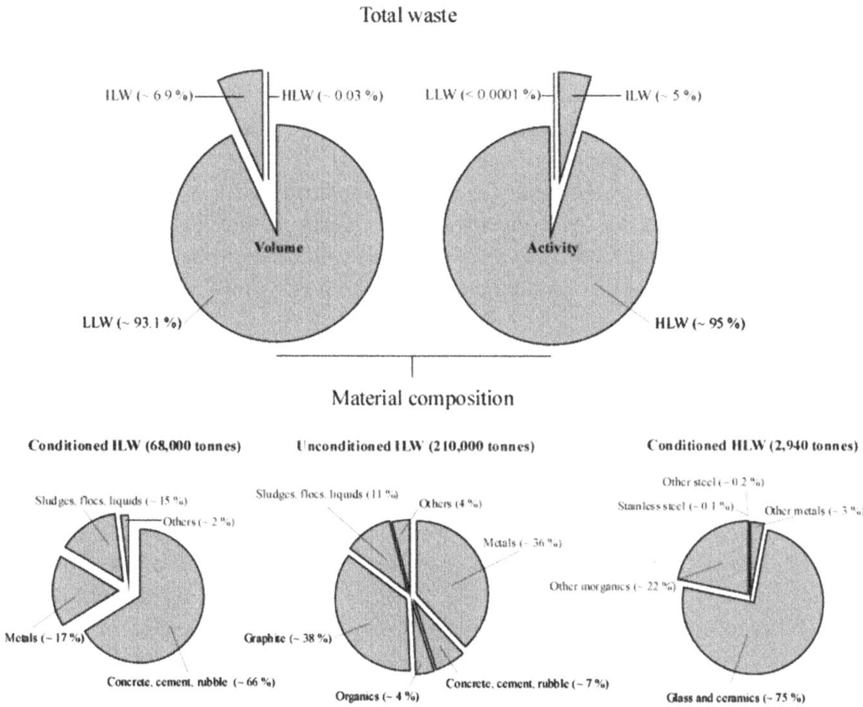

Figure 1 Volume, activity, and material contributions of existing UK HAW (data from the UK 2007 Radioactive Waste Inventory).[2]

more stable wasteform by glassifying or vitrifying it prior to storage and ultimately, disposal.[3] As extracted uranium and plutonium may have economic value, they are not currently classified as waste and are instead stored.

2.2 Intermediate Level Waste

Intermediate level wastes are "materials with radioactivity levels that exceed the upper limits for LLW but do not need heating to be taken into account in the design of storage or disposal facilities".[2] The majority of ILW arises during spent fuel reprocessing at the Sellafield site and consists of heterogeneous materials such as: magnesium alloy cladding that is stripped from Magnox fuels; steel, zircaloy and graphite components from AGR and PWR fuels; aqueous waste, sludges, flocs and organic materials from PUREX and radionuclide waste stream purification treatments; aqueous waste, flocs and filters from pond water treatments; and contaminated machinery from reprocessing operations (see Figure 1).[2] Most of the remaining ILW is produced at nuclear power stations where ILW principally arises during reactor operations and during subsequent decommissioning. Historical waste storage activities at a range of nuclear sites also give rise to ILW. As a consequence, the ILW waste stream has a highly heterogeneous and chemically challenging composition (see Figure 1).[2]

2.3 Low Level Waste

Low level waste is defined as material having a radioactive content between not exceeding 4 GBq tonne^{-1} of alpha, or 12 GBq tonne^{-1} beta/gamma activity.[2] Most LLWs consist of contaminated paper, plastic and metal products, and these are currently disposed of *via* shallow burial at the UK national LLW Repository Site, Drigg. However, a small proportion of LLW is unsuitable for shallow disposal due to the concentration of certain long lived radioisotopes (*e.g.* ^{237}Np, ^{239}Pu and ^{241}Am) in the waste,[2] as well as chemical incompatibility issues. These LLWs will be managed *via* geodisposal and treated in the same way as ILWs.[2]

2.4 Other Potential Wastes

Some existing radioactive materials not currently designated as wastes due to their potential economic value may be declared as waste in the future and thus may need to be managed through geological disposal.[1] Such items include spent fuel from UK nuclear reactors that is not already contracted for reprocessing; uranium and plutonium stocks from spent fuel reprocessing; uranium from fuel manufacturing; and non standard or "exotic" fuels from experimental and research reactors. These other potential wastes represent a significant increase in the activity for disposal. For example, as a result of past and currently planned reprocessing operations at Sellafield, the UK will have amassed approximately 100 tonnes of separated plutonium. Regardless of their eventual management route, these materials, in combination with the HLW discussed in section 2.1, represent exceptionally challenging wasteforms. For example, plutonium stocks potentially present major challenges for disposal due to both nuclear security and criticality issues. In addition, wastes from any new nuclear build and Ministry of Defence sources are also likely to be managed *via* geodisposal.[1]

3 Geological Disposal

3.1 The GDF Concept

Geological disposal involves the emplacement and isolation of HAWs in an underground repository (a GDF), housed deep inside a suitable rock formation (see Figure 2). For UK purposes, the definition of geological disposal is "burial underground (200–1000 m) of radioactive waste in a purpose built facility with no intention to retrieve".[1,4,5] Geological disposal facilities utilise a multiple barrier concept, whereby several engineered barriers are intended to work together with the host geology to contain and retard the radionuclides that are present in radioactive wastes. The components of a multiple barrier GDF typically include (see Table 1):

(i) *The waste form*. Wastes are conditioned (see section 3.3.5) prior to disposal to make them more stable. For example, highly active raffinate

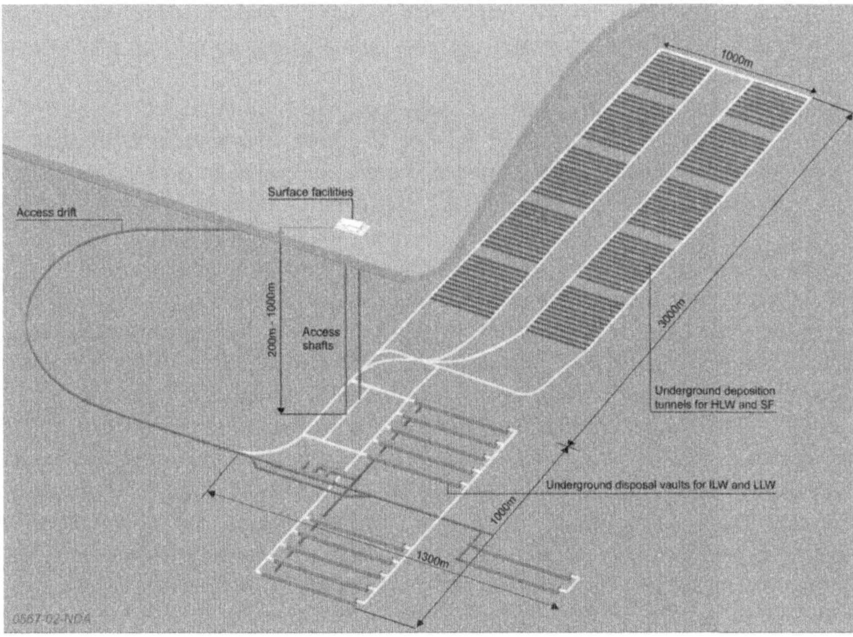

Figure 2 Schematic representation of a generic co-located geological disposal facility for HLW/spent fuel (SF) and ILW/LLW.[14] (Reproduced with permission from the UK NDA).

from nuclear fuel processing is vitrified to make it into an insoluble waste form. In the case of ILWs, grout encapsulation buffers the pH of the waste to hyperalkaline conditions under which a range of radiologically significant radionuclides (including the actinides) are predicted to be poorly soluble.

(ii) *The waste container*. The conditioned waste form is encapsulated in a container prior to disposal, creating a waste package. For example, grouted ILW is encapsulated in steel to provide mechanical stability. Furthermore, when the GDF evolves, steel corrosion creates reducing conditions that should retard the mobility of some radionuclides, particularly the actinides.

(iii) *Buffer materials*. Buffer materials are directly emplaced around the waste package. The materials used are chosen to provide beneficial functions, for example, to control the chemical or flow conditions in the repository during GDF evolution.

(iv) *Backfill*. Backfill is used to pack the GDF excavation tunnels, shafts, and drifts. The materials used must have the mechanical strength to support the GDF structure and are chosen to complement waste package conditioning and to allow further conditioning of the GDF to retard radionuclide mobility by pH, redox and/or flow control.

(v) *Sealing systems.* Highly impermeable sealing materials are required to control GDF groundwater ingress during construction and after waste emplacement.
(vi) *Geology.* The GDF host geology provides the final barrier for the repository. The geological barrier is likely to provide a number of beneficial functions, for example, it could support low groundwater flow or contain minerals and surfaces that sorb radionuclides from solution.

Once a GDF is closed, natural hydrochemical and biogeochemical processes will act to degrade the engineered structure and the multiple engineered barriers are expected to contain the waste for several thousand years.[1] After this engineered containment period, it is expected that groundwaters will have penetrated the backfill and canisters, and ultimately will have entered the waste packages, dissolving some fraction of the radionuclides. The partially degraded barriers will consist of a range of evolved mineral phases including iron oxides and aged cement phases, and for a cementitious repository, the pH is predicted to evolve from hyperalkaline to a more mildly alkaline state. Overall, the evolved system is likely to limit the mobilisation of radionuclides from the wasteform for several thousand to several tens of thousands of years. Nonetheless, with geological time radionuclides will transport to the surrounding geological environment, which will have been affected by the alkaline fluids from the GDF. In such instances, the radionuclides are likely to dilute and disperse, and sorption reactions with the surrounding rock and associated minerals and surfaces are intended to limit radionuclide transport into the biosphere to "acceptable levels". Clearly, a key challenge in successful GDF implementation is the associated communication of these uncertainties to the relevant stakeholder and public audiences. It is clear that high quality, independent, peer reviewed science is essential to allow full scrutiny of the safety case for geological disposal. It is also clear that communicating these scientific findings, coupled to explanation of the proposed GDF concept as it develops, in a clear and transparent way is likely to be pivotal in developing the "trust" between all players that is needed for movement toward GDF construction.

3.2 International Experience

3.2.1 Suitable Host Geologies

Globally, at least 39 countries have produced significant amounts of HAW and of these, 25 have chosen geological disposal as their long term HAW management pathway, and a further six have expressed a preference for geological disposal. However, whilst the current implementation pathway for geological disposal is relatively recent in the UK, several countries are more advanced in the implementation of their geological disposal strategies (see Table 1). Typically, these strategies have developed over several decades and all have included detailed geological characterisation programmes, often including construction of *in situ* rock testing laboratories to assess the

Table 1 Current overview of selected international HAW geological disposal concepts.

Country/Responsible body	*Current stage and Repository type*	*Site/Geology*	*Waste types managed by geodisposal*	*Waste package*	*Buffer/Backfill*
Sweden/SKB[6–8,11]	*Stage*: very advanced – inception in the early 1970s; thereafter, two suitable sites were identified; extensive site characterisation was conducted; an *in situ* testing facility constructed; engineering reference cases are well developed; material and host rock feasibility tests have been carried out; a final site has been chosen and permit applications are in progress. *Repository*: drifts leading to access tunnels with discrete (isolated) vertical boreholes in tunnel floor for waste emplacement. Discrete boreholes minimise rock fracturing. All waste will be disposed of in one facility; waste emplacement will be phased with	*Site*: potential site has been identified by SKB: Forsmark in the Östhammar area. *Geology*: site lies in crystalline geology. Initial site selection was confined to the stable Swedish Precambrian crystalline bedrock as other lithologies did not occur in sufficient quantities within Sweden. Repository will be constructed at a depth between 400–700 m where hydraulic conductivity is low and perceived glaciation risks are low.	Spent fuel. Long-lived ILW.	Spent fuel rods will be encased in copper canisters. Mechanical strength will be provided *via* speheroidal graphite cast iron inserts. The copper cladding will ensure corrosion protection for long time periods (1000's to 10 000's of years). ILW: conditioning is currently under review.	Spent fuel waste packages will be placed in isolated vertical boreholes. ILW emplacement is under review. The spent fuel waste package will be surrounded by highly compacted bentonite blocks/rings which will swell and seal the waste upon water ingress. Bentonite also offers good thermal conductivity, mechanical protection, and chemical stability, and poor hydraulic conductivity. Either natural Na bentonite of Wyoming type (MX-80) or Ca bentonite (deponite Ca-N) will be used. Two backfill materials are currently considered for the

Table 1 Continued.

Country/Responsible body	*Current stage and Repository type*	*Site/Geology*	*Waste types managed by geodisposal*	*Waste package*	*Buffer/Backfill*
	retrievability potential during early waste emplacement.				drifts and shafts: (1) pre compacted blocks of swelling clay with cavities filled with pellets of the same material, or (2) pre-compacted blocks made of bentonite and crushed rock (70 : 30 mixture), with crushed rock used to fill cavities. ILW: latest plans include stacking waste packages in caverns.
Finland/Posiva[21]	*Stage*: in construction – inception in the late 1970s; four suitable sites identified; site characterisation, environmental impact assessments, and public consultations conducted at all sites; final site chosen; *in situ* rock testing facility	*Site*: final disposal site has been chosen and ratified by Finnish government: Olkiluoto in Eurajoki. *Geology*: site lies in stable Precambrian crystalline geology (magmatic gneiss). Waste emplacement	Spent fuel.	As per the Swedish KBS-3 concept (detailed above).	As per the Swedish concept (detailed above).

	constructed. *Repository*: as per Sweden but with vertical or horizontal boreholes; retrievability considered possible at all stages of disposal.	at 420 m; hydraulic conductivity is low and groundwaters at this depth are generally devoid of O_2 (limiting corrosion).			
France/ ANDRA[6,11,22,23]	*Stage*: advanced – inception in 1991; a suitable host geology has been identified and characterisation is ongoing; engineering reference cases are well developed; *in situ* underground testing facility operational. *Repository*: horizontal or near horizontal boreholes off excavated drifts and shafts; all waste will be disposed of in one facility, with separate zones designated for different waste types; repository will be phased with retrievability considered.	*Site*: Meuse–Haute–Marne area of the Parisian basin at Bure has been identified as a potential area. *Geology*: Callovo-Oxfordian clay (a 130 m thick layer of indurated clay with low permeability and limited hydraulic exchange, that offers chemical and mechanical stability over long (geological) timescales. Cracks in clays are also self-sealing. Callovo-Oxfordian clay lies at approximately 400 m depth and is over- and underlain by limestone deposits.	HLW (vitrified waste from fuel reprocessing). Spent fuel (MOX and UOX) (note: spent fuels are not considered as waste in France but may be managed *via* geodisposal and thus are considered in the design process). Long-lived ILW.	HLW will be placed in unalloyed steel containers to limit water ingress during waste thermal stage. Spent fuels will be placed in a cast iron insert to provide mechanical strength, ensure sub criticality, and improve thermal transfer; the spent fuel array will then be encapsulated in an unalloyed steel container to prevent water ingress. ILW will be conditioned in bitumen or concrete, or will be compacted. Waste will be encapsulated in steel and placed in concrete overpacks. Waste packages will be stacked in tunnels/shafts.	HLW disposal boreholes will not use any buffer. Spent fuel boreholes will be lined with bentonite and steel. All boreholes will be sealed with a steel, bentonite and concrete plug. Concrete lined drifts will be backfilled with excavated clay and sealed with bentonite clay and concrete supporting structures. Shafts will be sealed with concrete and plugged with bentonite and excavated clay.

Table 1 Continued.

Country/Responsible body	*Current stage and Repository type*	*Site/Geology*	*Waste types managed by geodisposal*	*Waste package*	*Buffer/Backfill*
Belgium/ONDRAF-NIRAS[6,9–11]	*Stage*: advanced – inception in mid 1970s; two suitable host geologies have been identified; generic geodisposal concepts developed; engineering reference cases are well developed; *in situ* underground testing facility operational. *Repository*: waste inserted axially into dead-end disposal tunnels off an access tunnel; all waste disposed in one facility; repository phased with retrievability not considered.	*Site*: final site has not been decided but two suitable geologies have been identified. *Geologies*: boom clay formation and Ypresian clay formation. Boom clay (Tertiary age) is poorly indurated but has better mechanical strength than the Ypresian clay (Paleogene age). Both are argillaceous and of homogeneous composition, have low permeability and limited hydraulic exchange.	HLW (vitrified waste from fuel reprocessing). Long-lived ILW.	Vitrified HLW/spent fuel (contained in steel with silica glass frit used to fill voids) placed in carbon steel super-container. Ordinary Portland cement concrete used as buffer. ILW conditioned in bitumen or cement and encapsulated in 200 or 400 litre steel drums, drums placed in pre-fabricated monoliths. Voids filled with concrete.	HLW/spent fuel: void space between super-container and tunnel wall will be filled with cementitous material. Tunnels will be sealed with a concrete and clay plug. ILW: monoliths placed in concrete lined waste emplacement caverns/tunnels/galleries.
Switzerland/NAGRA[6,11,24]	*Stage*: very advanced – inception in the mid 1980s; thereafter, generic assessments included two rock types (crystalline and sedimentary); extensive	*Site*: no site has been identified. *Geology*: sedimentary – Opalinus clay (Jurassic age); clay layer is not thick (~100 m) but this	Spent fuel. HLW. ILW.	Vitrified HLW and spent fuel will be encased in carbon steel canisters. ILW: conditioned in concrete or bitumen; packed	HLW and spent fuel waste packages will be placed in horizontal disposal tunnels; HLW and spent fuel waste packages will be

	rock characterisation has been was conducted an *in situ* testing facilities have been constructed in both rock types; the sedimentary assessment (Opalinus clay/NAGRA) is presented here; material and host rock feasibility tests have been conducted. *Repository*: access tunnels leading to disposal tunnels; waste is emplaced axially; all waste will be disposed of in one facility; waste emplacement will be phased with retrievability considered.	will provide adequate (~40 m) coverage above and below the facility; the clay is homogeneous, and chemically and tectonically stable; the repository will be constructed at a depth of ~650 m.		in steel waste drums which are grouted into low-permeability waste emplacement packages.	surrounded by highly compacted bentonite clay; tunnels will be sealed with compound structures composed of bentonite, gravel, dry-stone (*e.g.*, granite or basalt), a bentonite and sand mixture, and a concrete plug. Bentonite chips and powdered bentonite will be used as mass backfill. ILW waste emplacement packages will be stacked in concrete lined tunnels. A gas permeable mortar will be used for backfill.
Germany/DBE Technology[6,11,25]	*Stage*: advanced – historically, salt rock was considered optimal for German HAW storage; final site (Government approved) selected in the late 1970s after	*Site*: Gorleben salt dome, NE Germany. *Geology*: salt rock (halite) – salt dome is homogeneous and thick (up to	Spent fuel (concept mainly developed for spent fuel). HLW (concept adapted to include HLW).	Spent fuel is encased in massive POLLUX containers (inner part stainless steel, outer part nodular cast	A buffer is not required for spent fuel/HLW (spent fuel alternative) concepts. A crushed salt backfill will

Table 1 Continued.

Country/Responsible body	*Current stage and Repository type*	*Site/Geology*	*Waste types managed by geodisposal*	*Waste package*	*Buffer/Backfill*
	directed state wide search; extensive site characterisation and testing carried out thereafter; engineering design concepts developed from the mid-1980s; the site and safety concepts are considered robust but a political moratorium was placed on the programme in 2000. *Repository*: shafts leading to drifts; waste is emplaced in drifts or in long vertical boreholes drilled through drift floors; waste emplacement is phased and retrievability is not considered but may be possible.	>1000 m), thus providing adequate coverage above and below the proposed facility; supports limited/no groundwater flow; is overlain by sedimentary rocks; is considered chemically and tectonically stable; the repository would be constructed at a depth of ~ 870 m. Salt rock exhibits plasticity under natural stresses (pressure, heat) and thus will seal waste in impermeable rock.		iron). Container is placed in drifts. HLW (and alternative for spent fuel) involves encapsulation in thin walled steel containers with emplacement in vertical boreholes. After emplacement and facility closure, the container provides no long-term function.	be used to fill drifts/boreholes. Disposal tunnels will be backfilled with crushed rock and sealed with mixed media (salt concrete/bitumen). This will limit groundwater penetration of the salt rock until creep seals the facility.

suitability of regional geologies. As a consequence, several generic host lithologies have been deemed suitable for HAW disposal (see Table 1). For example, the Belgian and French programmes have identified suitable clay formations; the Finnish and Swedish programmes have identified crystalline host rocks; the Germans have suggested potential evaporite deposits; and the Swiss have identified both crystalline and sedimentary formations (see review by Baldwin *et al.* 2008; ref. 6). Furthermore, Sweden has successfully advanced to the stage of choosing their likely geodisposal site, and Finland has begun pre-construction of their facility pending final government approval (see Table 1). Interestingly, the most successful national programmes have had public participation during site selection at the core of their implementation programmes. They also have less complicated nuclear legacies compared with that of the UK.

3.2.2 Engineering Approaches

Engineering considerations for multiple barrier structures have been the subject of extensive research, with design considerations varying between countries and according to waste type (see Table 1). For example, the latest Swedish design concept for spent fuel which has also been adopted by Finland (KBS-3 concept; see Table 1) favours waste encapsulation in corrosion-resistant copper canisters, with waste packages emplaced in isolated boreholes lined with bentonite, and with clay or bentonite blocks and crushed host rock used a backfill (see Table 1).[7,8] In contrast, the Belgian Spent fuel/HLW concept favours waste encapsulation in steel supercontainers, which are axially emplaced in disposal tunnels, with Ordinary Portland Cement (OPC)-based concrete used as a buffer and cement-based materials used as backfill (see Table 1).[9,10] Generally, international plans for ILW favour co-disposal alongside spent fuel and HLW, albeit in separate areas of the same repository (see Figure 2 and Table 1). Furthermore, ILW waste conditioning is typically in concrete/grout with encapsulation in steel and emplacement in boreholes or caverns that are backfilled with concrete, grout, clay and/or crushed host rock (see Table 1; review by Hicks 2008; ref. 11).

It is also worth mentioning that the USA currently hosts the world's only operational GDF, the Waste Isolation Pilot Plant (WIPP). This facility receives transuranic (TRU) wastes containing uranium, plutonium and other actinides, which are broadly similar to UK ILW. The site selection process for this facility began in the 1950s and waste emplacement commenced in 1999. WIPP is housed in stable salt rock formations, and design features include waste packaging in steel containers, waste emplacement in horizontal boreholes or shafts, and the use of MgO as a backfill material.[11] However, whilst WIPP has commenced geodisposal of US defence related wastes, the long term management pathway for US civil wastes is uncertain (and subject to current review) due to the US Department of Energy (DOE) recently withdrawing their license application for a HLW repository at Yucca Mountain, Nevada.[12]

3.3 Implementing the UK GDF

3.3.1 Historical Perspective, Public Consultation, Policy Decisions, and Responsibilities

Previous investigations concerning the geodisposal of UK ILW between the 1980s and 1990s were not successful and were effectively abandoned in 1997.[13] However, in 1999, the House of Lords Science and Technology Select Committee report on the status of UK radioactive waste[13] concluded that geological disposal was feasible and desirable but that the public should be consulted on future policy decisions. Accordingly, in 2001, the UK Government initiated the Managing Radioactive Waste Safely (MRWS) programme with a public consultation, to find the best practicable management solution for UK's HAWs. Following feedback from the consultation process, the Government commissioned the Committee for Radioactive Waste Management (CoRWM) to offer independent advice on the best HAW management pathways. In 2006, CoRWM submitted a range of recommendations to the Government, indicating a preference towards geological disposal, coupled with safe and secure interim storage, and a programme of ongoing research and development.[4] In response, the Government announced their plans for the long term management of HAWs to Parliament in October 2006.[5] The announcement accepted CoRWM's recommendation of geological disposal and the Government instigated a further period of consultation to investigate how geological disposal should proceed. After consultation, the Government White Paper *Managing Radioactive Waste Safely: A Framework for Implementing Geological Disposal* was published in 2008.[1] This document sets out the detailed policy and plans for geodisposal, and identifies the Nuclear Decommissioning Authority (NDA) as the body responsible for managing the delivery of UK geological disposal. In response, the NDA created the Radioactive Waste Management Directorate (RWMD) to facilitate this process. The MRWS White Paper also identified that: (i) the Government would retain responsibility for policy concerning geological disposal; (ii) independent regulators would oversee adherence to national and international statutory controls; and (iii) CoRWM would be retained to provide independent scrutiny and advice to the Government on geological disposal plans and programmes.

3.3.2 Guiding Principles and Timeline

UK GDF planning is in its infancy and subject to change, however, a number of important guiding principles for GDF implementation were outlined in the MRWS White Paper:[1]

(i) Site selection for GDF construction will be based upon community volunteerism and the siting process will take several years (see section 3.3.3).

(ii) The GDF will be tailored to the UK baseline inventory which is both large by volume and radioactivity, and complex in character due to the extended history of nuclear power generation and nuclear weapons production (see section 3.3.4).

(iii) The GDF design process will be informed by international experience and best practice (see section 3.2 and Table 1).

(iv) The Government currently favours the construction of a single GDF that is capable of housing all current and potential future HAWs and spent fuel (if this is declared as waste). The construction of separate GDFs at one or multiple sites (one for HLW/spent fuel and one for ILW) is also possible, but will have an increased cost and environmental impact compared to a single GDF.

(v) Economic and security considerations favour rendering GDF wastes as irretrievable in the long term, but planning, design and construction must be conducted in a way that does not exclude the option of a relatively extended period of retrievability pending a final decision.

(vi) As GDF implementation will take several decades, HAWs should be conditioned to increase stability (see section 3.3.5) and interim storage must be improved to ensure the safe containment of wastes prior to GDF emplacement (see section 3.3.6).

(vii) Further to the MRWS White Paper, the NDA-RWMD published a summary report *Geological Disposal: Steps towards Implementation*[14] that (i) outlined the preparatory work undertaken by the NDA in lieu of a final GDF site decision, and (ii) identified a prospective timeline for GDF implementation. Importantly, the preparatory work has involved the development of several GDF reference scenarios (including geological and engineering considerations) and this is discussed in section 3.3.7. The prospective timeline estimates initial GDF waste emplacement in 2040 and site operation over several decades during which waste will be monitored and could be potentially retrieved (see Figure 3).

3.3.3 Site Selection

The site selection process for the UK GDF(s) is outlined in the MRWS White Paper.[1] The process is based upon community volunteerism, which is coupled to published geological sub-surface screening criteria (SSSC) that provide an initial assessment of the geological suitability for hosting a GDF. Furthermore, underpinning the volunteerism approach is the principle of "participation

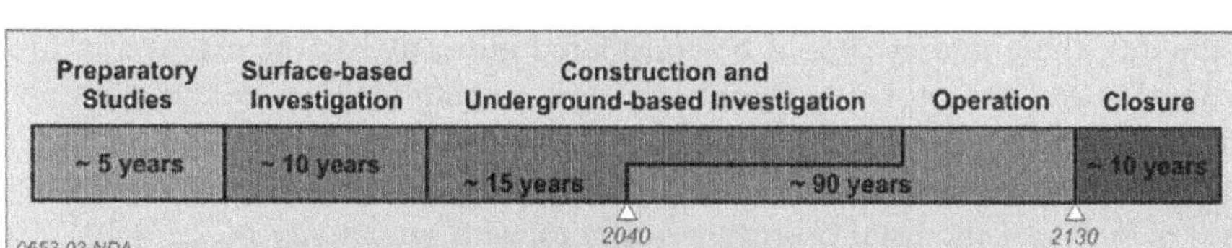

Figure 3 Representative UK geodisposal timeline indicating work phases.[14] (Reproduced with permission from the UK NDA).

without commitment"; here, host communities taking part in the GDF siting process can exercise a right of withdrawal up to the point where GDF underground observation and construction activities are due to begin. Currently, the UK government has extended an invitation to communities to submit expressions of interest in hosting the UK GDF. As of autumn 2010, two communities in west Cumbria have entered non-committal negotiations with the government and this process is ongoing. As part of this development, the British Geological Survey have been asked to apply the SSSC to the potential host communities, thus indentifying sites with unsuitable geology at an early stage.[1] The primary exclusion criteria are the presence of: (i) fossil fuel deposits; (ii) major areas of sub-surface waste or gas storage; (iii) potable water aquifers; (iv) extensive shallow (<500 m) permeable formations; or (v) complex hydrogeological environments (deep karstic formations or known source rocks for thermal springs).[1] If potential sites pass the SSSC, the host communities can then submit a formal "decision to participate" in the siting process. Thereafter, the site(s) will be subject to increased levels of scrutiny. This will initially involve desk based studies, and if appropriate, sub-surface investigations, until a decision can be made regarding the final host location. Importantly, the outcome of this process means that there can be no assumption that the GDF will be built in "ideal" geology; instead, the GDF host geology will depend upon the geological setting of the participating communities.

3.3.4 Inventory of Geodisposal Wastes

Planning for the UK GDF relies upon the accurate assessment of the types and quantities of HAWs destined for geodisposal. In response, the UK Government published a 'baseline inventory' of radioactive wastes in the 2008 MRWS White Paper.[1] The baseline inventory (see Table 2), is underpinned by the 2007 UK Radioactive Waste Inventory[2] and includes assessments of the likely volumes, activities, and material contributions of (i) current HAWs, (ii) future HAWs, and (iii) radioactive materials that are not currently classified as wastes (*e.g.* spent fuels and uranium and plutonium stockpiles). Accordingly, the baseline inventory defines the total volume (after packaging and conditioning) of geodisposal managed waste as ~480 000 m^3 and the activity $\sim 8.7 \times 10^{19}$ Bq. Of this, a minor volume (~0.3%) but a significant component of the activity (~41%) is designated as HLW (see Table 2). Spent fuels (that are currently not contracted for reprocessing) also comprise a minor volume (2.3%) but major activity (~52%) proportion of the inventory; however, along with HLW estimates, these figures should be considered indicative as the extent and lifespan of current reactors and spent fuel reprocessing facilities is uncertain and the treatment of spent fuel may change (see section 3.3.5). With these caveats in mind, the estimated volume of ILW destined for geological disposal is 364 000 m^3 (~76.3% of the total inventory volume), with an activity of 2.2×10^{18} Bq (2.5% of the total inventory radioactivity) (see Table 2). LLW is estimated to constitute ~3.6% of the potential waste volume, and <0.1% of the radioactivity (see Table 2). Lastly, the UK GDF generic design assessment must

Table 2 UK baseline of radioactive wastes destined for geodisposal.[1]

		Packaged volume		*Radioactivity (at 1st April 2040)*	
Materials	*Notes*	*Cubic metres*	*%*	*TBq*	*%*
HLW	a–c,e	1400	0.3	36 000 000	41.3
ILW	a,b,e	364 000	76.3	2 200 000	2.5
LLW	a,b,e	17 000	3.6	<100	0
Spent fuel	a,d,e	11 200	2.3	45 000 000	51.6
Plutonium	a,d,e	3300	0.7	4 000 000	4.6
Uranium	a,d,e	80 000	16.8	3000	0
Total	–	476 900	100	87 200 000	100

[a]Quantities of radioactive materials and wastes are consistent with the 2007 UK Radioactive Waste Inventory.[2]
[b]Packaging assumptions for HLW, ILW and LLW not suitable for disposal at the existing low-level waste repository are taken from the 2007 UK Radioactive Waste Inventory.[2] Note that they may change in the future.
[c]The HLW packaged volume may increase when the facility for disposing the canisters, in which the vitrified HLW is currently stored, has been implemented.
[d]Packaging assumptions for plutonium, uranium and spent nuclear fuel are taken from the 2005 CoRWM Baseline Inventory.[26] Note that they may change in the future.
[e]Radioactive data for wastes and materials were derived using the 2007 UK Radioactive Waste Inventory.[2] 2040 is the assumed start date for the geological disposal facility.
[f]It should be noted that at present the baseline inventory is based on UK inventory figures, and as such, currently contains waste expected to be managed under the Scottish Government's emerging policy for radioactive wastes.

consider the inclusion of radioactive wastes from any new build nuclear power stations.[1,15] However, the current baseline inventory does not include estimates of radioactive wastes arising from any new nuclear build program. Clearly, the development and operation of new UK reactors will lead to significant amounts of waste, and the government is thus committed to updating the baseline inventory as the GDF planning and any new nuclear build programmes proceed.[1] It is important to note that the HAWs (including the spent fuel) from the potential reactor designs submitted to the evolving new build program are consistent with a geodisposal management pathway.[16,17]

3.3.5 *Conditioning and Packaging of Geodisposal Wastes*

Conditioning is the immobilisation of radioactive waste in a suitable medium to produce a stable or solid waste form usually within a packaged container. The containerised waste can then be stored for several decades and ultimately will be disposed as the GDF is implemented. UK conditioning practices vary according to waste type:

(i) HLW: The highly active raffinate (HAR) solution from fuel reprocessing (see section 2.2) produces excessive heat and radioactivity and is highly unstable. Consequently, HAR is evaporated to reduce its volume to form highly active liquor (HAL). The HAL is then stored in water cooled tanks to permit heat dissipation and radioactive decay. After storage, the HAL is homogenised, immobilised, and conditioned in a vitrification plant. Here, the HAL is initially calcinated to remove water and nitrate, and then heated with crushed borosilicate glass to form molten glass.[3] The molten glass is then encapsulated and cooled in 150 litre stainless steel containers. This vitrified waste constitutes the high level waste form destined for geodisposal, but it is important to note that fuel reprocessing gives rise to large quantities of ILW (see section 2.3). As of 2007, approximately one third of existing UK HLW had been conditioned[2] and there is a commitment to reprocess all MAGNOX type fuel prior to closure of the Magnox reprocessing facility in ~2016 (ref. 18). The lifespan of the THORP reprocessing plant and hence the extent of oxide fuel reprocessing is uncertain; however, the NDA is contractually committed to reprocess approximately half of the UKs spent oxide fuel.[18] The fate of the remaining spent fuel is currently under consideration, but the NDA expect that a proportion may not undergo reprocessing and thus will be directly managed *via* geodisposal.[18] Finally, spent fuel from Sizewell B and from potential new build reactors, is currently considered unlikely to undergo reprocessing.[1,18]

(ii) Spent fuel: The eventual outcome of the UK policy toward oxide fuel reprocessing (see above), and decisions regarding any UK new nuclear build, will ultimately determine the quantity of spent fuel that is destined to be managed *via* geodisposal. As the UK does not currently declare spent fuel as waste, a bespoke UK geodisposal conditioning concept is not available. However, international best practice (see section 3.3.2 and Table 1) will likely inform future UK spent fuel strategies; indeed, the NDA explicitly consider the Swiss NAGRA SF disposal concept (see Table 1) in their latest geodisposal implementation document.[14]

(iii) ILW: For UK ILW waste forms (see section 2.3 and Figure 1), immobilisation in a cement based matrix and encapsulation in steel drums is the standard (existing) conditioning approach[2] and this reflects international practice (see Table 1). To date, approximately one fifth of UK ILW has been conditioned in this way[2] and pragmatically, the existence of these waste packages will likely inform future conditioning practices and UK GDF design; for example, a cavern system (see Table 1) is likely to be necessary to accommodate the large ILW volume, further, chemical compatibility issues between cementitous ILW and likely HLW/SF waste forms and buffer materials must also be considered. It is also important to note that a small but significant quantity of UK ILW has been conditioned in an organic polymer matrix[2] and that graphite from AGR reactors is a large volume, difficult ILW material.

3.3.6 *Interim Storage of Geodisposal Wastes*

The planning and construction of a GDF will take several decades. Consequently, safe interim storage of HAWs is an integral part of their long term management.[1,4] To ensure smooth transfer from storage to disposal facility, current interim storage plans must therefore be updated to meet a design principle that facilitates the storage of HAWs for 100 years or more in a manner that protects both people and the environment.[1] This represents a significant undertaking as large quantities of UK HAWs are stored in aging facilities. In addition, it is noteworthy that the current UK HAW stores are concentrated at the Sellafield site in west Cumbria, thus, any GDF implementation plan will need to consider transport of these materials to the eventual GDF location.

3.3.7 Reference Scenarios

As discussed, initial planning for UK GDF implementation is subject to significant uncertainty and at this stage flexibility in approach is essential. Particularly, a range of potential host geologies and repository designs must be considered at a generic level and any advanced case studies are presumptive and thus inappropriate at the current pre-site selection stage. Consequently, the NDA-RWMD has implemented a preparedness approach and has recently published a document that outlines the current UK GDF implementation strategy.[14] In this paper, a matrix of generic host settings reflecting typical, potentially suitable UK geologies and repository designs have been selected to demonstrate the viability of UK geodisposal, and better inform the conceptual design processes. The resulting host rocks where the GDF may be located include: higher strength rocks (typically crystalline igneous, metamorphic, or geologically old sedimentary rocks where fluid movement is supported through rock fractures *e.g.*, granite); lower strength sedimentary rocks (typically young sedimentary rocks where fluid movement is through pore spaces *e.g.*, clay); and evaporites (rocks that result from the evaporation of water containing salts *e.g.*, halite). Further, the covering rocks included in the generic geological assessment include host rocks to the surface, or sedimentary cover rocks. The resulting matrix of possible geological scenarios (see Table 3) indicates that all of the above potential geological combinations are possible in the UK, with the exception that that UK evaporite deposits do not extend to the surface.

Reflecting the geological considerations, the NDA have identified a matrix of illustrative GDF concepts (see Table 4) to inform their scoping work. These concepts reflect international experience (see Table 1), but also consider the existing UK ILW geodisposal concept. This relatively well developed concept[19] stems from the UK's unsuccessful attempt to implement geodisposal in the 1990s (see section 3.3.1). In developing this concept, it was assumed the GDF

Table 3 Matrix of considered host and cover rocks possible in the UK.[14]

	Host Rocks		
Cover rocks	*Higher strength rocks*	*Lower strength sedimentary rocks*	*Evaporites*
Host rocks to surface	Possible	Possible	Not possible
Sedimentary cover rocks	Possible	Possible	Possible

Table 4 Matrix of geodisposal concepts considered by the NDA.[14]

	Illustrative geological disposal concept examples	
Host rock	*ILW/LLW*	*HLW/Spent fuel*
Higher strength rocks[a]	UK ILW/LLW concept (NDA, UK)	KBS-3 Concept (SKB, Sweden)
Lower strength sedimentary rocks[b]	Opalinus clay Concept (NAGRA, Switzerland)	Opalinus Clay concept (NAGRA, Switzerland)
Evaporites[c]	WIPP bedded salt concept (US-DOE, USA)	Gorleben Salt Dome Concept (DBE-Technology, Germany)

[a]The NDA selected the UK ILW/LLW concept and the Swedish KBS-3V concept for spent fuel due to the availability of information on the concepts for the UK context.
[b]The Opalinus clay concept for HLW, spent fuel and ILW was chosen by the NDA because a recent OECD Nuclear Energy Agency review regarded the NAGRA (Switzerland) assessment as "state of the art" with regard to knowledge. However, the NDA detail that they will also consider aspects of the French (ANDRA) and Belgian (ONDRAS/NIRAS) concepts.
[c]The NDA state that the WIPP assessment was chosen for ILW due to the wealth of information available on this licensed, operating facility. Likewise, the HLW/spent fuel decision (DBE-Technology, German) was based upon the level of concept information available.

would be housed at several hundred metres depth in a geological setting that comprised high permeability sedimentary rocks overlying low permeability sedimentary rocks, with the GDF housed in low permeability hard rock (see Hicks 2008; ref. 11). In this model, after conditioning (in cement and steel), the waste would be stored then emplaced in a GDF as the facility becomes available. After several decades to centuries, when waste emplacement had occurred and when the operational lifetime of the GDF was complete, the GDF would be sealed with a cementitous backfill and abandoned. At this point, the sub-surface GDF environment would then contain conditioned and packaged waste, along with significant volumes of cement and structural iron, and the sub-surface would resaturate with groundwater. The cementitous waste packaging and backfill were designed on resaturation to maintain high pH conditions promoting hydrolysis of metal ions (including the actinides) and thus minimising radionuclide solubility. In addition, the use of iron metal in the facility (in both packaging and engineering structures) is intended to promote strongly reducing conditions, again limiting the solubility of some radionuclides.

4 Environmental Chemistry Research Challenges in Geological Disposal

The global legacy of radioactive wastes combined with the urgent need for low carbon power generation means that we are now at a pivotal time for the implementation of geological disposal facilities for radioactive wastes. In the UK, we have a large and complex legacy of materials from over half a century of nuclear power plant operation and weapon production. Indeed, the fact that the UK was the first nation to implement nuclear power generation, that we have been active in creating a bespoke research and development programme in nuclear power, and that we have always had an extensive reprocessing programme, means that we have a waste legacy that is both diverse and highly challenging. With this background in mind, it is useful to define our view on some of the key environmental chemistry challenges facing successful HAW geodisposal:

(i) For ILW (assuming a cementitous waste form and repository environment), there are challenges in understanding and predicting the corrosion and evolution of the waste form and host rock environment during storage and disposal.
(ii) Microbial processes can affect the solubility of radionuclides;[20] however, the fundamental biogeochemistry of radionuclides (especially the long lived radionuclides of importance to geological disposal including the transuranics) is poorly defined, especially under geological disposal settings. The presence of electron donors such as organic matter in the waste, H_2 from radiolysis of water and anaerobic corrosion of iron metal, and iron metal itself within the GDF, combined with the increasing recognition that "extremophile" microbes can tolerate high pH conditions and extreme radiation fluxes, means that understanding the whole biogeochemistry of a evolving GDF will be critical in underpinning the safety case.
(iii) For HLW/spent fuel, there are challenges in co-locating the waste within the same GDF as ILW as it is likely that a large fraction of ILW will be grouted and backfilled with cement. This hyperalkaline ILW concept is chemically incompatible with international HLW and spent fuel concepts although separate "sub-chambers" for ILW and HLW that take into account the predicted hydrogeology of the subsurface may offer a route forward.
(iv) There is a poor understanding of the long term waste form performance of UK specific fuels such as advanced gas cooled reactor oxide fuels.
(v) Colloids have the potential to influence radionuclide transport both within and outside of an evolved GDF environment.
(vi) It is important to note that computational modelling, from an atomistic scale to a regional scale, is required to underpin the predictive model(s) for GDF performance assessment. This is because the scientific community cannot hope to perform experiments on all the systems of interest (particularly with highly radiotoxic transuranics) over all of the time and spatial scales of relevance.
(vii) Research with radionuclides presents a significant safety hazard and the UK infrastructure and capability to perform these experiments has been eroded in the last decades. As a consequence, successful GDF implementation will require innovation, collaboration, and investment across a range of scientific disciplines.

In conclusion, it is important to realise that the safety case for the successful implementation of a UK GDF will involve environmental chemistry challenges that encompass a range of scales from sub micro-second reactivity to millions of years; from molecular scale understanding of biogeochemical processes to field scale transport modelling; and from essentially pure radionuclide phases such as UO_2 fuel to sub 10^{-12} molar concentrations of radionuclides that are likely to be transported from the GDF into the host rock. Implementation of the UK GDF is a unique and exciting challenge and it requires innovation across the scientific disciplines and a commitment to engage with the wider public to deliver the reward of safe and credible management of the nuclear legacy. Thus our view is that the management of radioactive wastes forms a real focus for environmental chemistry research that will enable both management of a difficult and challenging legacy and potentially allow consideration of a new fleet of nuclear reactors to power the future energy needs of the UK.

Acknowledgements

We thank Prof. Francis Livens for helpful discussions and for proof reading the manuscript. We acknowledge support from NERC for ongoing research on nuclear legacy management, particularly grant numbers NE/H007768/1, NE/D00473X/1, NE/D005361/1.

References

1. Department for Environment, Food and Rural Affairs, Department for Business, Enterprise and Regulatory Reform, Welsh Assembly Government, and Department of the Environment Northern Ireland, *Managing Radioactive Waste Safely: A Framework for Implementing Geological Disposal*, 2008.
2. Nuclear Decommissioning Authority and Department for Environment, Food, and Rural Affairs, *UK Radioactive Waste Inventory 2007 Main Report*, 2008.
3. C. Sharrad, L. M. Harwood and F. R. Livens, *Nuclear Power and the Environment*, eds. R. E. Hester, J. M. Harrison, Royal Society of Chemistry, Cambridge, UK, 2011, pp 39–55.
4. Committee on Radioactive Waste Management, *Managing our Radioactive Waste Safely – CoRWM's Recommendations to Government*, CoRWM Document 700, 2006.
5. UK Government, *Response to the Report and Recommendations from the Committee on Radioactive Waste Management (CoRWM)*, 2006, www.defra.gov.uk/environment/radioactivity/waste/pdf/corwm-govresponse.pdf (accessed October 2010).
6. T. D. Baldwin, N. A. Chapman and F. B. Neall, *Geological Disposal Options for High Level Waste and Spent Fuel*, Galson Sciences Limited report to NDA-RWMD, 2008.
7. SKBF/KBS, *Report on the Final Storage of Spent Nuclear Fuel – KBS-3*, 1983.
8. SKB website and associated information materials, http://www.skb.se (accessed October 2010).

9. ONDRAF/NIRAS, *Safety Assessment and Feasibility Interim Report*. NIROND 2001–06–E–December 2001.
10. ONDRAF/NIRAS website and associated information materials, www.nirond.be (accessed October 2010).
11. T. W. Hicks, T. D. Baldwin, P. J. Hooker, P. J. Richardson, N. A. Chapman, I. G. McKinley and F. B. Neall, *Concepts for the Geological Disposal of Intermediate-Level Radioactive Waste*, Galson Sciences Limited report to NDA-RWMD, 2008.
12. United States of America Department of Energy, *Motion to Withdraw License Request for HLW Repository at Yucca Mountain*, Nevada, 2010, http://www.energy.gov/news/documents/DOE_Motion_to_Withdraw.pdf (accessed Oct 2010).
13. UK House of Lords Science and Technology Select Committee Report on the Management of Nuclear Waste, 1999, http://www.parliament.the-stationery-office.co.uk/pa/ld199899/ldselect/ldsctech/41/4102.htm (accessed October 2010).
14. Nuclear Decommissioning Authority, *Geological Disposal: Steps Toward Implementation*, NDA report number NDA/RWMD/013, 2010.
15. Department for Business, Enterprise and Regulatory, *Reform Meeting the Energy Challenge: A White Paper on Nuclear Power*, 2008.
16. Nuclear Decommissioning Authority, *Generic Design Assessment: Summary of Disposability Assessment for Wastes and Spent Fuel arising from Operation of the UK EPR*, 2009.
17. Nuclear Decommissioning Authority, *Generic Design Assessment: Summary of Disposability Assessment for Wastes and Spent Fuel arising from Operation of the Westinghouse AP1000*, 2009.
18. Nuclear Decommissioning Authority, *Draft Strategy Published for Consultation*, 2010.
19. NIREX, *Generic Repository Design*, Nirex Report N/077, 2003.
20. J. R. Lloyd and J. Renshaw, Microbial transformations of radionuclides: fundamental mechanisms and biogeochemical implications in *Metal Ions in Biological Systems*, ed. A. Sigel, H. Sigel and R. K. O. Sigel, CRC Press, Boca Raton, FL, USA, 2005, vol. 44, 205–240.
21. POSIVA website and associated information materials, http://www.posiva.fi/en (accessed October 2010).
22. ANDRA website and associated information materials, http://www.andra.fr (accessed October 2010).
23. ANDRA, *Dossier 2005 Argile. Synthesis: Evaluation of the Feasibility of a Geological Repository in an Argillaceous Formation, Meuse/Haute-Marne Site*, ANDRA, France, 2005.
24. NAGRA, *Project Opalinus Clay Safety Report*, NAGRA technical document NTB02-05, 2002.
25. DBE Technology website and associated information materials, www.dbetec.de/en/about-dbe-tec/publications (accessed October 2010).
26. Committee on Radioactive Waste Management, *CoRWM's Radioactive Waste and Materials Inventory*, CoRWM Document 1279, 2005.

Pathways of Radioactive Substances in the Environment

JOANNA C. RENSHAW,* STEPHANIE HANDLEY-SIDHU AND DIANA R. BROOKSHAW

ABSTRACT

The release and transport of radionuclides in the environment is a subject of great public concern. The primary sources of radionuclides in the environment are nuclear weapons testing and production, and the processes associated with the nuclear fuel cycle. Whilst nuclear weapons tests have been the main source of atmospheric contamination, resulting in global, low-level contamination, sites associated with weapon production and the nuclear fuel cycle can have localised high levels of contamination, and the spread of this contamination *via* aquatic pathways represents a significant environmental problem. Migration in the atmosphere will depend on the nature of the radioactive material and the prevailing meteorological conditions. Within surface water and groundwater environments, transport will be controlled by physical processes such as advection and the biogeochemical conditions in the system. In systems with significant flow, advection will be the dominant transport process, but as hydraulic conductivity decreases, chemical processes and conditions become increasingly important in controlling radionuclide migration. Factors such as solution phase chemistry (*e.g.* ionic strength and ligand concentrations), E_h and the nature of mineral phases in the system have a critical effect on radionuclide speciation, controlling partitioning between solution and solid phases and hence migration. Understanding

*Corresponding author

Issues in Environmental Science and Technology, 32
Nuclear Power and the Environment
Edited by R.E. Hester and R.M. Harrison

Published by the Royal Society of Chemistry, www.rsc.org

the complex interplay between these parameters is essential for predicting radionuclide behaviour and migration in the environment.

1 Introduction

The primary sources of radioactive contamination are the nuclear weapons and nuclear energy programmes over the past 70 years.[1,2] In addition, other sources of radionuclides in the environment include the accidental release of radioactive material used in medicine and industry; the release of naturally occurring radionuclides through other industrial processes (primarily mining and mineral processing); and, more recently, the use of depleted uranium in weapons.[2] In the environment, radionuclides can be transported *via* the atmosphere or aquatic systems, either as surface waters or through terrestrial systems in groundwater. Figure 1 summarises the main sources of radionuclides, the environmental pathways and the key processes controlling radionuclide migration. Whilst nuclear weapons tests have been the primary source of global, low-level contamination *via* the atmosphere, the other sources of radionuclides can cause localised high levels of contamination, and the spread of this contamination *via* aquatic pathways, and possible uptake into the food chain, represents a real environmental risk. In this chapter, the main pathways for radionuclide migration in the environment and factors controlling migration are reviewed, focusing mainly on geochemical factors controlling transport in aquatic systems.

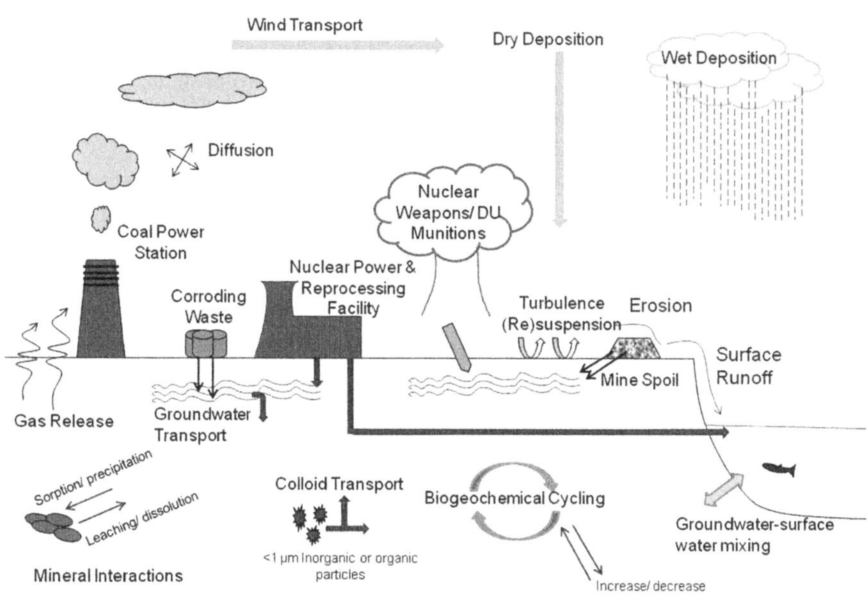

Figure 1 Summary of main sources of radionuclides, the environmental pathways and the key processes controlling radionuclide migration.

2 Sources of Radionuclides in the Environment

2.1 Nuclear Weapons

Nuclear weapons tests account for a significant proportion of the total activity released into the environment and historically are the major source of radionuclides in the atmosphere. An estimated 2×10^8 TBq of radioactivity have been released into the atmosphere as a result of nuclear weapons testing;[3] Table 1 lists the radionuclides produced and released by atmospheric nuclear tests.[4] Most of the radionuclides released were short-lived, and so atmospheric levels of radioactivity have declined sharply from their peak in the 1960s; further decline in levels of radioactivity will be much slower, as the remaining activity is predominantly due to long-lived ^{14}C.[4] Fallout from atmospheric weapons testing will also cause contamination of surface water and terrestrial environments. Fallout can either be local (within a few 100 km of the test site), regional (up to several thousand km from the site) or global, and the spread of fallout will depend on the altitude and latitude of the explosion and the explosive yield.[5,6]

Although much of the contamination arising from nuclear weapons testing is widely dispersed and at low levels, there are considerable levels of activity at test and production sites. In the USA, there are >70 million m^3 of contaminated soil and > 1800 million m^3 of contaminated water at Department of Energy facilities used for weapons production.[7] At the Mayak Production Association in the Chelyabinsk region, Russia, weapons-grade plutonium was produced for ~40 years and significant levels of contamination exist at the site and the surrounding area from both production and accidental discharges.[8,9] Approximately 10^5 TBq of radioactivity, as liquid waste, were discharged from the site into the Techa River between 1949 and 1956, with most of the released radioactivity associated with $^{89+90}Sr$ (20.4%), ^{137}Cs (12.2%), rare earth isotopes (26.8%), ^{95}Zr–^{95}Nb (13.6%) and ruthenium isotopes (25.9%).[9,10] At the same site ~7.4×10^4 TBq of radioactivity were released as a result of a high level radioactive liquid waste tank exploding, causing the contamination of

Table 1 Radionuclides produced and globally dispersed in atmospheric nuclear tests.[4]

Radionuclide	*Half-life*	*Global Release (PBq)*	*Radionuclide*	*Half-life*	*Global Release (PBq)*
H-3	12.33 y	186000	Sb-125	2.76 y	741
C-14	5730 y	213	I-131	8.02 d	675000
Mn-54	312.3 d	3980	Ba-140	12.75 d	759000
Fe-56	2.73 y	1530	Ce-141	32.5 d	263000
Sr-89	50.53 d	117000	Ce-144	284.9 d	307000
Sr-90	28.78 y	622	Cs-137	30.07 d	948
Y-91	58.51 d	120000	Pu-239	24110 y	6.52
Zr-95	64.02 d	148000	Pu-240	6583 y	4.35
Ru-103	39.26 d	247000	Pu-241	14.35 y	142
Ru-106	373.6 d	12200			

~20 000 km^2 at concentrations >4000 Bq m^{-2}.[8,11] Underground weapons testing has caused contamination of the subsurface with tritium, fission and activation products and actinides.[2,12] At the Nevada Test site, the primary location for nuclear weapons tests in the USA, ~ 1×10^7 TBq of radioactivity was released into the subsurface during 828 tests.[13] The decay-corrected radionuclide inventory as of 1992 (the year of the last test) is 4.86×10^6 TBq, with the most significant amounts of radioactivity arising from ^{3}H, ^{137}Cs, ^{90}Sr, $^{241+239}Pu$, ^{85}Kr, $^{152+154}Eu$ and ^{151}Sm.[14] The inventory will change, however, as short-lived radionuclides decay and daughter radionuclides appear; with time, the remaining radionuclide inventory in the subsurface will be dominated by long-lived radionuclides such as uranium, plutonium, neptunium and americium.[14]

2.2 *Nuclear Fuel Cycle*

The other major source of radioactive waste and contamination is the nuclear fuel cycle.[2,7] By volume, the largest source of contamination arises from uranium mining and milling.[15] Uranium mining has produced an estimated 937×10^6 m^3 of tailings, with activities ranging from <1 to >100 Bq g^{-1}.[16] The waste contains not only uranium, but also uranium decay products, including radon, a radioactive gas. Although current tailings are well maintained, there are many old abandoned sites, particularly in eastern European countries and the former Soviet Union that require remediation.[4,16]

Contamination also arises from the handling and reprocessing of spent nuclear fuel (from either civil or military programmes) and can cause high localised levels of contamination.[1,17] Release to the environment can arise from authorised discharges to the atmosphere and to surface and groundwater (see Tables 2 and 3), accidental release and leakage of storage tanks.[18–21] At the Sellafield site, UK, authorised discharges to atmosphere and sea have occurred for over 40 years.[2] Historically, the major sources of liquid effluent for discharge (*via* pipelines into the Irish Sea) were process liquors from reprocessing and fuel storage pond water; discharges for selected radionuclides from 1952 to 1992 are shown in Figure 2.[22] The level of activity discharged to sea peaked in the mid to late 1970s, and in most cases has been declining ever since.

Leakage from storage facilities can also cause significant localised contamination. At the Hanford site (a former plutonium production facility) in the

Table 2 Airborne releases of carbon-14, tritium, iodine-129, krypton-85, total β/γ emitters and total α-emitters, from European spent fuel reprocessing sites from 1999 to 2003.[21] Values given are in GBq per annum. Total β/γ excludes ^{14}C, ^{3}H and ^{129}I.

Year	*C-14*	*H-3*	*I-129*	*Kr-85*	*Total β/γ*	*Total α*
1999	2719	3.2×10^5	33.47	3.86×10^8	3.64	0.22
2000	2676	2.85×10^5	32.06	3.08×10^8	3.00	0.13
2001	972	3.05×10^5	24.42	3.31×10^8	2.88	0.11
2002	857	3.17×10^5	31.49	3.46×10^8	2.84	0.07
2003	737	4.41×10^5	22.26	3.72×10^8	3.91	0.17

Table 3 Liquid releases of carbon-14, tritium, total β/γ emitters, total α-emitters and uranium from European spent fuel reprocessing sites from 1999 to 2003.[21] Values given are in GBq per annum, except for uranium which is in kg per annum. Total β/γ excludes ^{14}C and ^{3}H.

Year	*C-14*	*H-3*	*Total β/γ*	*Total α*	*Uranium*
1999	1.57×10^4	1.54×10^7	1.48×10^5	217.75	545.86
2000	1.31×10^4	1.28×10^7	1.15×10^5	206.16	614.70
2001	1.67×10^4	1.22×10^7	1.63×10^5	279.30	392.84
2002	2.09×10^4	1.52×10^7	1.69×10^5	495.22	444.64
2003	2.57×10^4	1.58×10^7	1.25×10^5	503.55	488.38

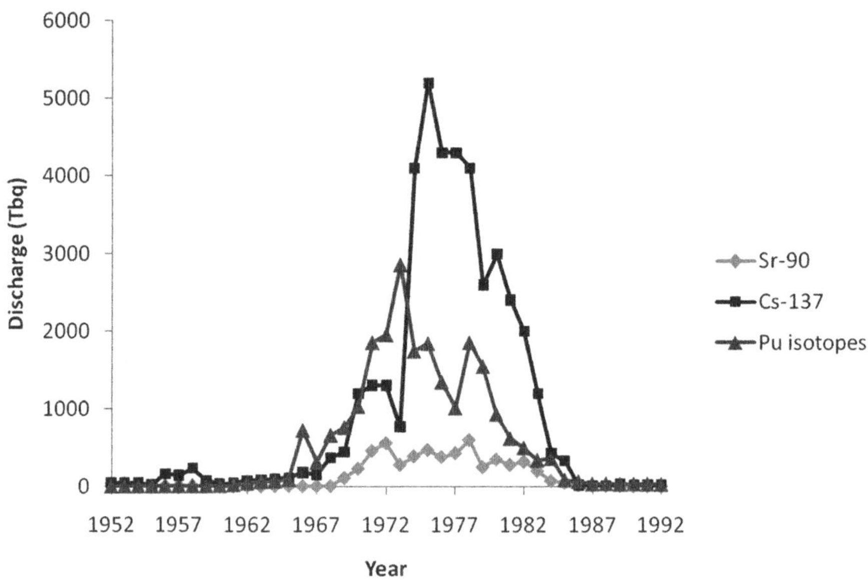

Figure 2 Discharges to the Irish Sea from the Sellafield site for selected radionuclides, 1952–1992.[22] Plutonium isotopes are ^{238}Pu, ^{239}Pu and ^{241}Pu.

USA, an estimated 570 m^3 of waste containing 3.7×10^4 TBq of radioactivity has been released into the subsurface from leaking underground storage tanks.[7]

Releases to the atmosphere arising from the nuclear fuel cycle are lower than from nuclear weapons, but can arise from several stages of the cycle. Uranium mining and milling releases radon gas and windborne dispersion of waste materials can also spread contamination.[23,24]

2.3 Depleted Uranium

In addition to nuclear weapons testing, another military source of radionuclides in the environment has been the use of depleted uranium (DU) in recent conflicts. DU, a by-product of nuclear fuel enrichment, is used in anti-tank penetrators due to its high density, self-sharpening and pyrophoric properties.[25] During Gulf War I, approximately 320 tonnes of DU munitions were used by the US forces.[26] The amount of DU fired by the US forces during the recent 2003 Iraq War has not yet been yet been disclosed but speculative figures range between 170–1700 tonnes.[27] Although DU is less radioactive than natural uranium, health effects can arise from both radiological and chemical toxicity.[26,28]

2.4 Naturally Occurring Radioactive Material

Naturally occurring radioactive isotopes are either primordial (*e.g.* ^{40}K, ^{238}U or ^{232}Th; present from the creation of earth), including their decay products, or cosmogenic (formed by cosmic rays).[4,29] Primordial radionuclides and the decay products of ^{238}U and ^{232}Th represent the more significant problem with regard to environmental contamination,[30] but concentrations vary significantly in the environment and depend on local geology.[4] Migration of and exposure to naturally occurring radionuclides can be significantly enhanced by industrial activities, such as mining and mineral processing, in particular in production of phosphate,[31,32] oil production and combustion of coal (which contains trace quantities of radionuclide) in power stations.[31,33,34] The radionuclide concentrations in different types of coals range from 12–435 Bq kg^{-1} for ^{238}U; 21–309 Bq kg^{-1} for ^{226}Ra; 7.5–56 Bq kg^{-1} for ^{232}Th and 6–398 Bq kg^{-1} for ^{40}K.[34] When coal fuel is burnt in power plants the ashes generated are enriched in metals and radionuclides. The amount of ash released into the atmosphere from coal-fired power plants can vary from 10% in an old plant, to 0.5 % in modern emission-controlled power plants.[35] In addition, coal burning also releases radon into the atmosphere.

Although globally the release of naturally occurring radionuclides through industrial activities is a relatively minor source of contamination, compared to civil and military nuclear programmes, it can still result in local elevated levels of contamination. Flues *et al.* (2002)[35] found a one- to three-fold increase in the natural radionuclide concentrations (^{232}Th, ^{226}Ra and ^{210}Pb) within a 1 km distance of a 10 MWe coal-fired power plant, and elevated radon levels have been reported in and around coal power plants but the dose is less than the recommended occupational exposure limit.[36]

2.5 Accidental Release

Accidental release of radionuclides from nuclear facilities or from other sources (*e.g.* industrial or medical) is a less significant source of radioactive contamination in the environment and the largest releases have been due to the accidents at Chernobyl and Fukushima.[4] At Chernobyl, 1.76×10^{18} Bq of ^{131}I and 8.5×10 Bq of ^{137}Cs were released into the atmosphere;[4] at Fukushima preliminary calculations estimate that 1.5×10^{17} Bq of ^{131}I and 1.2×10^{16} Bq of ^{137}Cs were released between 11th March and 5th April 2011 (ref: Japan Nuclear Safety Commission, press release 12th April 2011). Nuclear accidents are discussed in detail in Chapter 3. From other sources, one of the most serious incidents occurred at Goiania, Brazil, in 1987. A radiation source from a cancer therapy machine was scavenged from an abandoned clinic and sold to scrap dealers, who opened the source, containing 50.9 TBq of ^{137}Cs in the form of a luminescent powder.[37,38] The material attracted a great deal of interest and was distributed to family and friends of the scrap dealers. As a result, 249 people were contaminated, with 21 suffering acute radiation sickness or radiodermatitis; four people died and another six were in a serious condition. Seven sites covering 5000 m^2 in Goiania were found to be highly contaminated and clean up involved the demolition of houses and the construction of repositories for the waste.[38]

3 Environmental Chemistry of Key Contaminants

Radioactive waste contains a wide range of isotopes, but many are either short-lived radionuclides or are stable (*i.e.* non-radioactive) isotopes. The radionuclides that are problematic environmentally are those that are long-lived, have a high activity, are present in relatively large quantities and/or are bioavailable. Details of some key radionuclide contaminants are given in Table 4. All are present in significant quantities in the environment from nuclear weapons testing or the nuclear fuel cycle, except for ^{60}Co; uranium and radon also occur naturally. Cobalt-60 is widely used in medical and industrial applications requiring a radiation source, and is created by neutron activation of ^{59}Co.

Several of these radionuclides are bioavailable. Radon exists as a gas and inhalation of radon can cause cancer; it is the second leading cause of lung cancer in the USA.[39] Strontium(II), an analogue for Ca^{2+} can be accumulated in bone, whilst Cs^+ is analogous to K^+ and so can be transported into cells *via* the K^+ transport mechanisms; both ^{99}Tc and ^{129}I can be accumulated in the thyroid gland.[40,41]

The environmental fate of radionuclides is controlled by a number of factors; these will be discussed in detail in sections 4 and 5. However, the oxidation state

Table 4 Key radionuclide contaminants.

	Oxidation states	*Key Isotopes*	*Half-life*	*Major decay mode*
Fission products				
Strontium	+2	^{90}Sr	29.1 y	beta
Technetium	+4, +7	^{99}Tc	2.15×10^5 y	beta
Iodine	−1, 0, +5	^{129}I	1.57×10^7 y	beta, gamma
Caesium	+1	^{137}Cs	30.17 y	gamma
Actinides				
Uranium	+3, +4, +5, +6	^{238}U	4.47×10^9 y	alpha
Neptunium	+3, +4, +5, +6, +7	^{237}Np	2.14×10^6 y	alpha
Plutonium	+3, +4, +5, +6, +7	^{238}Pu	87.7 y	alpha
		^{239}Pu	2.41×10^4 y	alpha
		^{240}Pu	6.55×10^5 y	alpha
		^{241}Pu	14.4 y	beta
Americium	+3, +4, +5, +6, +7	^{241}Am	432.7 y	alpha
Other				
Cobalt	+2, +3	^{60}Co	5.271 y	beta, gamma
Radon	0	^{222}Rn	3.8 d	alpha

of the radionuclide will have a significant impact on its chemical behaviour, transport and bioavailability, particularly for redox-sensitive radionuclides. As can be seen from Table 4, the actinide elements can exist in a range of oxidation states, leading to fairly complicated chemical behaviour. The most stable oxidation state of uranium in the environment is U(VI), as UO_2^{2+}, but it is also stable as U(IV) under reducing conditions.[42] The most stable and dominant oxidation states of neptunium and plutonium in the environment are +V (as NpO_2^+) and +IV, respectively. However, in the environment, neptunium can also exist in the +IV and +VI oxidation states, whilst plutonium can also be present in +III and +V oxidation states.[43] The higher, environmentally stable oxidation states (V, VI) of the actinide elements tend to be more soluble and therefore more mobile, whilst An(IV) species (An = actinide), with a high charge/radius ratio, are prone to hydrolysis and polymerisation, forming colloids and precipitates, and readily sorb to mineral surfaces.[1,44]

Of the fission products listed, technetium and iodine are redox active, but caesium and strontium have only one stable oxidation state each: Cs^+ and Sr^{2+}. Therefore changes in the redox environment do not directly affect the chemistry of caesium and strontium, but the environmental behaviour and bioavailability of Cs^+ and Sr^{2+} do relate to their oxidation state. The low charge density on Cs^+ means that it is only weakly complexed by ligands and tends to bond *via* electrostatic interactions rather than covalent bonding. It is also highly soluble and so is mobile in the environment, with interactions with mineral phases being the dominant mechanism of retardation.[44] Like Cs^+, Sr^{2+} does not complex strongly to ligands and tends to be soluble in the environment, but it can co-precipitate with calcium sulfate or carbonate.[44,45] The mobility of technetium is primarily controlled by the oxidation state, with two stable states found in the environment: +VII and +IV. Under aerobic conditions, technetium will exist in the +VII oxidation state, as TcO_4^-, and in this form it is highly soluble and mobile. Under more reducing conditions, Tc(IV) is stable and will tend to exist as insoluble TcO_2.[2] For iodine, the most important oxidation states in the environment are: −I, 0 and +V. In aqueous environments, +V (as IO_3^-) and −I (as I^-) are the dominant forms, but in soils, iodine can be mostly present as organic species.[1,2] The redox chemistry of cobalt is relatively simple, with just two stable oxidations states: +II and +III. Co(II) is the dominant oxidation state in solution, as it tends to be more soluble than Co(III), but Co(III) can be stabilised and mobilised by certain ligands (see section 5.3).[46,47]

4 Processes and Factors affecting Radionuclide Transport in the Atmosphere

Radionuclides may enter the atmosphere as gas, aerosol or particulate matter. The transport of suspended radionuclides is dependent on particle size; larger particles will settle and deposit faster than smaller particles. Following the release into the atmosphere, the dispersion of radionuclides is mainly controlled by meteorological conditions (*i.e.* winds, turbulence, advection and wet and dry precipitation), radioactive decay and diffusion.

For atmospheric nuclear weapons tests, transport of the radioactive debris will depend on the height and yield of the explosion, the nature of the debris, the location of the test site and prevailing meteorological conditions. Refractory radionuclides, such as plutonium, ^{95}Zr and ^{144}Ce, are released mainly in particulate form,[5,48] and so will tend to be deposited more rapidly, and be less widely dispersed, than more volatile radionuclides, such as ^{137}Cs and ^{131}I.[49] During testing, radioactive debris will be injected into the atmosphere at different heights, and this will depend primarily on the height of the test and the explosive yield; low yield tests will tend to release debris into the troposphere, with the quantity of radioactive material released into the stratosphere increasing with yield.[50] For tests conducted near the surface, it is estimated that around 50% of the debris is deposited locally or regionally, with the remainder more widely dispersed.[5,49] Debris released into the troposphere (the lowest level of the atmosphere) can be transported up to several thousand kilometres from the test site over 1–2 weeks, as a result of the turbulent air movements that occur there.[49,50] Removal of particulate debris from the troposphere is mainly caused by precipitation but dry deposition of radionuclides can also occur.[50] Radioactive debris released into the stratosphere remains in the atmosphere for much longer periods of time (>1 year) than material released at lower altitudes, and so will be dispersed over a much greater area, with precipitation the main mechanism for deposition.[6,10] As a result, global radioactive contamination arising from deposition of material from the stratosphere will consist of longer-lived radionuclides, compared to local and regional contamination.[6,49] Simon *et al.* (2004)[6] investigated the geographical distribution in the USA of radionuclide fallout arising from tests at the Nevada Test site (NTS) and global fallout. The distribution of radioactive debris from the low yield tests at the NTS depended on the wind patterns and local rainfall events at the time of the test, but in general the highest levels of deposition were in the region immediately east of the site. With global fallout, higher levels were deposited in the eastern and mid-western regions than the south-western states, reflecting the relative levels of precipitation in these regions.

More localised atmospheric transport of radionuclides occurs with the use of DU weapons and uranium mining and milling. When a DU munitions hits its target, an estimated 10%–35% (maximum of 70%) of the DU mass is converted into aerosol, with most of the dust particles <5 µm.[26] The transport of DU particles will depend on particle properties (*i.e.*, size and density) and on prevailing meteorological conditions.[51] Surveys of the post-conflict zone in Kosovo and Bosnia and Herzegovina reported DU contamination up to 200 m away from the point of impact.[52,53] Lloyd *et al.* (2009)[54,55] investigated the dispersion of aerosols formed during the combustion of waste metal at a uranium and DU processing factory in Colonie (NY, USA). The distribution of the DU aerosol was controlled by prevailing winds, with DU contamination found up to 600 m from the factory. It has been estimated that at least 3.4 tonnes of uranium was deposited within 1 km of the factory.[56] Resuspension of DU dust has also been found to occur by wind or human disturbance.

From uranium mining and milling, radon gas will be released into the atmosphere, but in arid climates, windborne dispersion of fine radioactive particulate wastes can also be a problem.[16,57] Lottermoser and Ashley (2006)[58] investigated the physical dispersion of radioactive waste from a rehabilitated uranium mine in South Australia. Under the semiarid conditions at this site, there had been significant wind dispersion of radioactive particulates from the site. Around the main tailings storage facility, tailings material up to 10 cm thick was spread up to 80 m from the source in the northeast and southeast sides, reflecting the prevailing wind directions at the site. Around this source an area of 1 km^2 had uranium concentrations >100 ppm, with another 2 km^2 contaminated with 10–100 ppm of uranium. Radon, generated in the subsurface or in waste materials, is mainly released into the atmosphere by diffusion, but advection caused by wind and changes in barometric pressure can also play a role and mining activities will enhance rates of release into the atmosphere.[24,59]

Radioactive materials released into the atmosphere from the accident at the Fukushima Nuclear plant were detected globally but at very low levels. Monitoring undertaken by the Preparatory Commission for the Comprehensive Nuclear-Test-Ban Treaty Organisation (http://www.ctbto.org/press-centre/highlights/2011/fukushima-related-measurements-by-the-ctbto/fukushima-related-measurements-by-the-ctbto-page-1/) detected traces in eastern Russia on 14th March 2011, three days after the earthquake and tsunami that damaged the reactors. Radiation was detected on the west coast of USA by 16th March and all across the northern hemisphere 15 days after the accident. The equator acts as a dividing line between the northern and southern air masses, and so the dispersal of radioactive materials was initially limited to the northern hemisphere; however, by 13th April, radiation from Fukushima had spread to the southern hemisphere.

5 Processes and Factors affecting Radionuclide Transport in Aquatic Systems

The behaviour and mobility of any radionuclide depends on its chemical speciation,[60] which will control properties such as solubility and reactivity with respect to surfaces.[15] Chemical speciation is critically dependent on the biogeochemical conditions; factors such as pH, E_h, the presence of complexants and the nature of mineral surfaces present in the system, and the interplay between these factors, will all affect the partitioning of radionuclides between the solid and solution phases.[15,61,62] Figure 3 illustrates the key geochemical processes controlling radionuclide speciation and some of the factors affecting these processes. Processes such as sorption and (co-)precipitation may retard radionuclide migration, whilst dissolution, complexation and colloid formation may enhance migration by retaining the radionuclide in the solution phase.

Waters in the environment have varying biogeochemical signatures. Surface waters can be subdivided into fresh water and saline environments, with areas of mixing (brackish or estuarine zones). Within the fresh water environments, rivers and streams are likely to be of neutral pH, low ionic strength and oxidizing conditions. In such systems, within significant flow, advection (transport

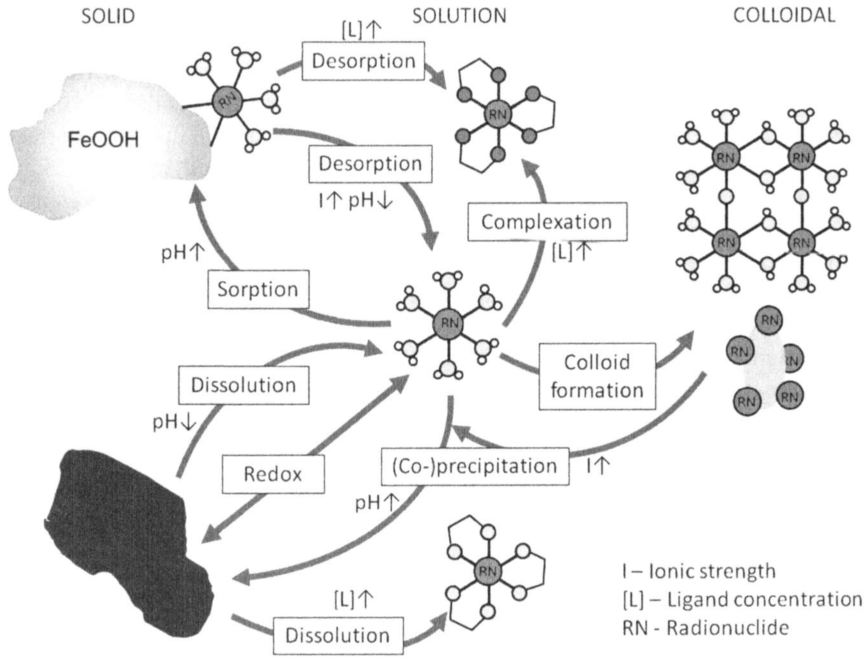

Figure 3 Summary of the key geochemical processes controlling radionuclide speciation and some of the factors affecting these processes.

of a solute due to the overall water flow) is the dominant transport process. However, as hydraulic conductivity of the system decreases, such as in ground water systems, advection has less impact on solute migration,[63] and chemical gradient-driven processes and diffusion processes become increasingly more important.[64] Within lakes and reservoirs or estuarine environments, mixing can be limited and zones of different pH and E_h conditions can develop.[43,65] In such environments, regions of more reducing conditions can promote retardation processes, such as microbial reduction of redox-sensitive radionuclides, which can limit radionuclide migration. In the subsurface, radionuclides are transported as solutes by groundwater flow, but the prevailing biogeochemical conditions will significantly affect migration. Groundwater environments are typically reducing due to the low oxygen penetration in such zones.

5.1 Sorption to Mineral Surfaces

Retardation of radionuclides in the subsurface primarily results from their interactions with minerals.[61] Sorption of radionuclides to mineral surfaces is controlled by the structure and charge at the mineral surface.[42] There are a number of mechanisms by which radionuclides can interact with the mineral phase, including ion exchange, chemisorption and physisorption.[45] With ion exchange, the sorbing ion exchanges with another similarly charged ion within the mineral structure. For Cs^+, cation exchange with clay minerals is one of the key processes controlling the mobility of the ion.[44]. When the sorbing cation binds through covalent bonding to the mineral surface, forming inner sphere complexes, the reaction is termed chemisorptions.[45,66] Bonding occurs only at specific sites on the mineral surface and the strength of binding depends on the metal ion. For example U(VI) can bind to iron oxyhydroxides (goethite and hydrated ferric oxide) and micas (muscovite and chlorite) through the formation of inner sphere complexes at the mineral surface.[66–68] Radionuclides also interact with the mineral surface through weak van der Waals' forces (physisorption),[69] forming an outer sphere complex. Such bonding is relatively weak and so radionuclides are readily desorbed from the surface by small variations in the geochemistry of the environment. This interaction is typical of Sr^{2+} with many mineral surfaces including ferrihydrite,[66] bacteriogenic iron oxides,[70] kaolinite[71] and calcite.[72]

Irregularities on the mineral surface, such as kink and step sites or etch pits, tend to be more reactive than other parts of the crystal surface and may therefore be the preferred site of adsorption[66] or microbially mediated dissolution of the mineral and any related processes.[73,74] The steric environment of the adsorption site in combination with the chemical composition of the ligands at the surface affects the affinity of the site for particular radionuclides. Micaceous minerals like illite,[75] montmorillonite and vermiculite[76] and biotite[73] have a high affinity for Cs^+ at their edge and step sites. Caesium(I) can diffuse into the interlayers of these sheet silicate minerals over time, making the adsorption irreversible.[77]

Sorption processes are affected by the biogeochemistry of the solute.[61,78] There is typically a strong dependence on pH, with most mineral surfaces being most efficient sorbents at circumneutral pH.[79] In more acidic environments, the large concentration of H^+ ions in solution causes protonation of mineral surfaces, altering their overall surface charge. In alkaline environments complexation of cationic radionuclides by hydroxyl ions (OH^-) decreases the positive charge of the cation. Both of these effects lower the electrostatic affinity between the radionuclide and mineral surface and so reduce sorption.

The ionic strength and cation concentration in the solution phase will also affect adsorption to mineral surfaces. Increasing ionic strength will reduce the activities of the radionuclides in solution and will also alter the effective charge at the mineral surface, in both cases decreasing complexation to the mineral surface.[62] Changes to the ionic strength of the solution phase can also cause desorption of radionuclides. Standring *et al.*[9] investigated the remobilisation potential of ^{137}Cs, ^{60}Co, ^{99}Tc and ^{90}Sr associated with sediments within Reservoir 10 along the Techa River system at the Mayak site. The sorption of caesium and technetium to sediments within the Techa River system was found to be relatively irreversible, however significant proportions of the sediment-bound strontium and cobalt could be remobilised upon mixing with fresh water or seawater. The desorption effect was significantly increased in the presence of seawater; this effect was attributed to the higher pH and ionic strength of the seawater.

Sorption properties may be modified by the presence of coatings on the surface of the mineral but the effect will depend on the radionuclide.[78] Adding an aluminium coating to illite, kaolinite and montmorillonite surfaces enhanced Sr^{2+} sorption compared to uncoated surfaces.[80] Coatings of the minerals with humic substances did not affect Sr^{2+} sorption significantly, but decreased Cs^{+} sorption compared to uncoated surfaces, and the magnitude of the effect was different for each clay mineral (strongest for illite and weakest for kaolinite), reflecting the specificity of adsorption systems.[80] Biological coatings, such as bacterial biofilms, on mineral surfaces may also modify the reactivity of the mineral surfaces by masking existing sorption sites. Anderson *et al.* (2007; ref. 81) found that the presence of a biofilm over a granite "fracture surface" reduced adsorption of Am(III), Pu(IV) and Np(V) and interpreted this as the biofilm preventing adsorption by decreasing the diffusion of the radionuclides near the mineral surface.

5.2 Redox Reactions

For redox-sensitive radionuclides such as uranium, technetium, neptunium and plutonium, oxidation state is one of the primary controls on mobility, affecting precipitation, complexation, sorption and colloid formation behaviour. The dominant oxidation states for some key radionuclides at pH 7 are shown in Figure 4. Microbial metabolism can drive a wide range of redox transformations, utilising a succession of terminal electron acceptors (TEA), including some redox-active radionuclides, for the oxidation of organic matter. The amount of

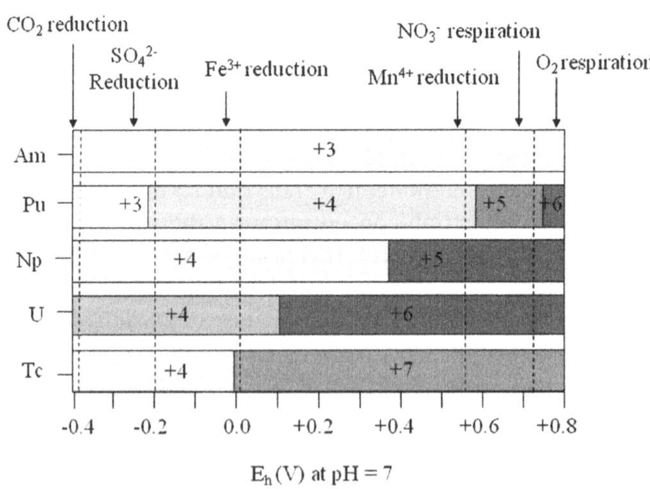

Figure 4 Expected dominant oxidation states as a function of E_h for radionuclides in 0.01M NaCl aqueous solution, pH 7 and equilibrated atmospheric CO_2. (Adapted from Morris and Raiswell).[17] Technetium data using artificial groundwater at pH 7 and equilibrated atmospheric CO_2 adapted from Hu *et al.*[143]

energy gained from the use of each TEA influences the rate and sequence of TEA utilisation. The classical TEA sequence is reduction of: O_2, NO_3^- Mn(IV), Fe(III), SO_4^{2-}, followed finally by methanogenesis,[17] but radionuclides such as uranium and technetium can also be used as TEAs. Under Fe(III) and SO_4^{2-} reducing conditions, U(VI) (as UO_2^{2+}) and Tc(VII) (as TcO_4^-) can be reduced by a wide range of microorganisms to less mobile U(IV) and Tc(IV), respectively.[1,82,83] There have also have been a few limited studies reporting microbial reduction of neptunium and Pu.[84–85] The Fe(III)-reducing bacteria *Geobacter sulfurreducens* and *Shewanella oneidensis* have been reported to slowly reduce Pu(IV), as amorphous $Pu(OH)_4$, to Pu(III); for *S. oneidensis,* the rate of reduction was increased by the presence of riboflavin as an endogenous redox mediator.[84] *Shewanella oneidensis* and a mixed consortium of sulfate-reducing bacteria have been found to reduced soluble NpO_2^+ to insoluble Np(IV).[85,86]

In addition to direct microbial reduction, the oxidation state of redox-active radionuclides will also be influenced by presence of microbially-generated redox-active species, reactive mineral phases and microbial alteration of mineral phases. Redox-active ions exposed at the surface of a mineral, such as sulfur or iron in mackinawite (FeS) can reduce an adsorbed radionuclide such as U(VI) and Tc (VII).[87] This changes the controls on subsequent remobilization processes, and makes oxidation the dominant re-suspension pathway rather than the presence of competing cations in solution or pH fluctuations. Livens *et al.*[87] found that the reduced uranium was readily reoxidised and desorbed upon introduction of oxygen. Reduced iron sediments within a soil profile can also immobilise Tc(VII) by similar surface-mediated reduction to Tc(IV).[88] In contrast to uranium, however, technetium does not remobilise as readily with oxygen when it is in association with mackinawite.[89]

Bacteria can reduce transition metals (notably iron and manganese) locked within mineral structures[90,91] and this could alter the characteristics of the reactive surface, generate new reactive mineral phases or release redox-active species into solution.[15,74] In particular, iron-bearing minerals can play a crucial role as mediators between microbial anaerobic respiration and redox sensitive radionuclides. Ferrous iron released by microbial reduction of iron-bearing phases can react with a number of ligands and phases present in solution to form a range of new iron phases: oxyhydroxides (magnetite, goethite),[92] carbonates (siderite) or phosphates (vivianite)[93] and others. The mineral phases formed in the environment can be hard to predict, but are likely to be dominated by carbonates and hydroxides due to the abundance of those ligands in solution. Wildung *et al.*[88] investigated technetium reduction in shallow aquifer sediments from the US Atlantic Coastal Plane. The primary control on the reduction of Tc(VII) was the amount of readily extractable (and so more reactive) Fe(II) present in the sediments. Other studies have also found that Tc(VII) can be reduced to Tc(IV), as TcO_2 by biogenic Fe(II) , with TcO_2 associated with the biogenic Fe(II) mineral phase.[94,95] Upon reoxidation of reduced sediments there can be significant reoxidation and remobilization of technetium, but it is dependent on the nature of the oxidant. When sediments are reoxidised with air, a significant (50–80%) fraction of the reduced and immobilised technetium can be reoxidised and remobilised; however, when the oxidant present is nitrate, there is much more limited (<10%) reoxidation of technetium.[96,97]

Bioreduction may also lead to dissolution of the mineral phase and loss of sorption sites.[98] Bacteria can reduce iron within a number of iron oxyhydroxides of varying crystallinity: hematite, goethite, lepidocrocite and schwertmannite,[99,100] and even micas such as biotite,[101] smectite[102,103] and illite.[99,104] This can cause release and remobilization of radionuclides which have been adsorbed or incorporated into the mineral phase. Langley *et al.* (2009)[105] investigated the impact of microbes on strontium sorbed to bacteriogenic iron oxides. Microbial reduction of the ferric iron within the iron oxides remobilised strontium (increased its concentration in solution), most likely due to loss of sorption sites. The authors suggest, however, that in a natural system remobilised strontium would be transported upwards by advection and recaptured within newly-formed bacteriogenic iron oxides near the surface of the water body, once again retarding its transport.

Under circumneutral conditions, abiotic or biotic oxidation of Fe(II) or Mn(II) leads to the formation of new oxyhydroxide phases.[106,107] These secondary mineral phases are characterised by large surface area and small crystal size, and have very high sorption capacity.[107,106] At low concentrations, oxidised uranium has been reported to form an inner sphere complex on biomineralising manganese oxides.[108] At high concentrations U(VI) was sequestered very efficiently and was incorporated into the oxide structure. The large cation caused distortion of the manganese oxide lattice and the formation of a mineral with tunnel-like structures. The results highlight the significance of the solution chemistry during mineral formation and the sequence of biomineralisation processes and the presence of radionuclides in solution.

5.3 Complexation Reactions

In any aqueous environment, cations will be complexed, either by water molecules (hydration) or by other ligands present. The natural environment contains a range of common ligands, such as CO_3^{2-}, OH^-, Cl^- and natural organic matter that can complex radionuclides. In addition, synthetic organic ligands can also be present as co-contaminants. For actinides the strength of inorganic ligand complexation decreases in the order: CO_3^{2-}, $OH^- > HPO_4^{3-}$, F^-, $SO_4^{2-} > NO_3^-$, Cl^-.[109] Actinides interact with anionic ligand species by strong ionic bonding and so the complexation strength is related to the actinide charge; therefore the trend of complexation is: $An^{4+} > AnO_2^{2+} > An^{3+} > AnO_2^+$.[110] In natural systems the most important inorganic ligand is carbonate; it is ubiquitous in the environment, with concentrations ranging from 10^{-5} M in surface waters to 10^{-2} M in groundwater.[64] Carbonates can form stable, negatively charged complexes with actinides. These complexes have a lower affinity for negatively charged mineral surfaces and so the complexed radionuclide will tend to remain in the solution phase, therefore increasing mobility in the environment. For example U(VI) has a high affinity for carbonate ligands, forming stable complexes such as $UO_2(CO_3)_2^{2-}$ or

$UO_2(CO_3)_3^{4-}$ as shown in equations (1) and (2).[111]

$$UO^{2+}_{2(aq)} + 2HCO_3^- \rightarrow UO_2(CO_3)^{2-}_{2(aq)} + 2H^+ \quad (1)$$

$$UO_2(CO_3)^{2-}_{2(aq)} + HCO_3^- \rightarrow UO_2(CO_3)^{4-}_{3(aq)} + H^+ \quad (2)$$

The speciation of dissolved uranium in the presence of inorganic carbon is summarised in Figure 5. In oxidising and mildly reducing environments at pH >6, the negatively charged $UO_2(CO_3)_2^{2-}$ or $UO_2(CO_3)_3^{4-}$ dominate, between pH 5– 6 the less mobile $UO_2(CO_3)$ species dominate and below pH 5 the uranyl ion (UO_2^{2+}) dominates.[112]

The presence of natural and synthetic organics may also increase the mobility of radionuclides. Natural humic substances are ubiquitous in the environment and concentrations can range from <1 to >200mg l^{-1} in wetlands.[113] Humic substances are composed of three fractions: the humin fraction is insoluble under all pH conditions; humic acids are insoluble at ≤pH 2, and fulvic acids are soluble under all pH conditions.[114] Humics and fulvics are important natural ligands that are able to complex with radionuclides and may increase mobility. Humic complexation is effected by pH; with increasing pH there is an

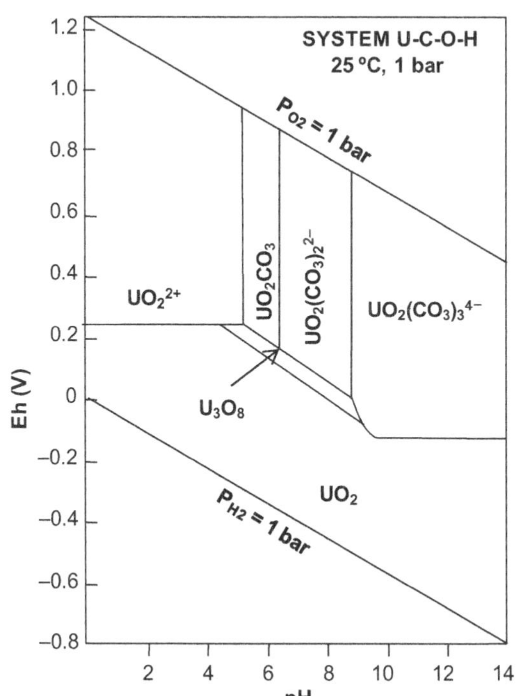

Figure 5 Redrawn E_h–pH diagram showing the U–C–O–H system. Assumed activities for dissolved species are: $U = 10^{-8,-6}$, $C = 10^{-3}$. (Adapted from Brookins,[112] with kind permission from Springer Science + Business Media: E_h – pH *Diagrams for Geochemistry, Uranium,* 1988, p154, Fig. 88).

increase in the ionisation of humic functional groups (*e.g.* carboxylic and phenolic), thus increasing the humic complexation strength.[115] The stability of humic complexes depends on the oxidation state of the complexed metal. The stability of radionuclide–humic complexes decrease in the order U(IV) > Th (IV) > Am(III) > Eu(III) > U(VI) > Co(II) > Sr(II), as a result of strength of electrostatic interaction between radionuclide and functional group.[42,116] Actinides–humic complexes can modify radionuclide oxidation states; the mediated reduction of Np(VI) to Np(IV) and Pu(VI) to Pu(IV) has been reported to occur.[117] Complexation by humics also depends on the presence of other complexants. Moulin and Moulin[118] investigated the effect of humics on actinide migration under conditions relevant to nuclear waste disposal (pH ≤ 7). Only An(III) were found to be complexed by humics, whilst An(V) and An(VI) were complexed with carbonates or hydroxide.

Organic acids can also be present in the environment as a result of microbial activity. These ligands can sequester cations from mineral surfaces or stabilise cations in the solution phase.[119] The acids released during microbial metabolic processes are also potent mineral dissolution agents.[119] Most minerals are stable at circumneutral pH and dissolve in the presence of acid[101,120] releasing adsorbed or incorporated contaminants.

Organic complexing agents used in the processing of nuclear fuel or during decontamination (*e.g.*, nitrilotriacetic acid (NTA), ethylenediaminetetraacetic acid (EDTA) and citric acid[121]) can be present as co-contaminants with radionuclides; for example the 149 radioactive waste storage tanks at the US Department of Energy contain an estimated 83 metric tons of EDTA.[122] Complexation of radionuclides by organic ligands can reduce the reduction potential of the metal and several studies have found that bacterial reduction of radionuclides increases in the presence of organic complexants. Laboratory studies showed only minor reduction of Pu(IV), present as amorphous $Pu(OH)_4$ to Pu(III) by *Shewanella oneidensis* and *Geobacter metallireducens*.[123] However, Pu(IV)–EDTA complexes were rapidly reduced to the more mobile Pu(III)–EDTA by the same bacteria.[123] In another study, the solubilisation of PuO_2 by Fe-reducing bacteria was increased in the presence of NTA by approximately 90%; the proposed mechanism was reductive dissolution of Pu(IV) to Pu(III).[124] The co-disposal of radioactive ^{60}Co(II) and EDTA is also a concern. In the environment the Co(II)–EDTA complexes can be oxidised by manganese(IV) and iron(III) oxide minerals to the more stable and mobile Co(III)–EDTA.[47] Although metal-reducing bacteria can reduce Co(III)–EDTA back to the less mobile Co(II)–EDTA, in a natural system re-oxidation by oxide minerals may occur.[125,126]

Co-contaminants can also stabilise radionuclides in the solution phase and therefore enhance migration in the environment. AlMahamid *et al.* (1996)[127] investigated complexation of Pu(III, IV, V and VI) with NTA and EDTA. The predominant oxidation state at pH 5 to 8 was Pu(IV); Pu(III) was oxidised and Pu(V/VI) were reduced under these conditions. Critically, the presence of NTA and EDTA stabilised Pu(IV) in solution. In another study, the co-contaminant citrate was found to decrease the adsorption of U(VI) onto iron-rich sand.[128] The dominant mechanism was thought to be chemical alteration of the sand surface by citrate, decreasing the adsorption of U(VI), but aqueous complexation of U(VI) by citrate may also be significant.[128]

5.4 (Co-)Precipitation

At low concentrations, sorption to mineral surfaces will be the dominant mechanism retarding radiouclide migration in subsurface environments. However, radionuclides can also be removed from the solution phase through co-precipitation as new mineral phases are formed and, at higher radionuclide/ligand concentrations, by precipitation. At lower concentrations, co-precipitation will be the more important process.[45] For the actinides, hydrolysis can occur for all oxidation states and can lead to precipitation from solution.[129] In particular, An(IV) has a strong tendency to hydrolyse.[110]

In natural waters, key mineral phases for (co-)precipitation will include carbonate and iron (oxy)hydroxide mineral phases. Parkman *et al.* (1998)[72] investigated the interaction of Sr^{2+} with calcite and found that at higher (≥ 0.3 mM) concentrations Sr^{2+} was precipitated as strontianite on the calcite surface, perhaps as a result of the existing mineral phase providing nucleation sites and higher localised carbonate concentrations at the calcite surface–water interface. NpO_2^+ has also been reported to co-precipitate with calcite.[130] Co-precipitation appeared to be independent of pH, even though pH did affect Np(V) solution speciation. Extended X-ray absorption fine structure (EXAFS) analysis suggested NpO_2^+ was structurally incorporated into the calcite lattice, with Np and the axial O atoms substituting for one Ca^{2+} and two CO_3^- groups, respectively. Co-precipitation of UO_2^{2+} with calcite is reported to be limited,[130] but natural aged calcite has been found with uranyl incorporated into the lattice[131,132] and one study investigating uranium contamination in the Aral Sea suggests co-precipitation with calcite and gypsum is the dominant mechanism for U removal from the solution phase.[133] U(VI) can also co-precipitate with iron oxides, with X-ray absorption spectroscopy (XAS) suggesting that it is incorporated into the oxide lattice as uranate (containing U–O single bonds and no axial U–O double bonds) rather than uranyl.[134]

5.5 Colloidal Transport

Colloids are particles that are <1 μm in at least one dimension, and with a high surface area, that remain suspended in the water column.[135,136] Colloids containing radionuclides can form either through condensation of particular radionuclide species by a hydrolytic or precipitation process, degradation of nuclear waste (intrinsic colloids), or through sorption of radionuclides onto colloids of other (inorganic or organic) material, for example, iron oxyhydroxides or humic acids (extrinsic or carrier colloids).[44,109,137]

Colloidal transport of radionuclide will be affected by the geochemical and physical properties of the water system.[135,138] Geochemical conditions will affect radionuclide sorption to colloids (as with sorption to mineral surfaces), colloid formation and colloid stability.[44,45] For example, high ionic strength can increase colloid coagulation, causing precipitation of the colloids from the water column.[139] Colloidal transport can be retarded by colloid deposition at solid–water and air–water interfaces, and by straining in pore systems, whilst shear and hydration forces can mobilise colloids.[135,139] However, binding to colloids and/or the formation of colloids by radionuclides can have a significant effect on radionuclide transport, with colloid-mediated transport often being more rapid than solution phase.[135,140] In studies investigating plutonium migration in groundwater at the Mayak site, Russia, plutonium was found to be both in solution and colloid-associated at distances up to 2.15 km; further afield (up to 3.9 km), 70–90 mol% Pu was associated with 1–1.5 nm colloids, suggesting a key role for colloid-facilitated transport to the far-field environment.[141] Mori *et al.* (2003)[142] investigated the effect of bentonite colloids on the transport of radionuclides through granodiorite at the Grimsel test site, Switzerland. In the absence of bentonite colloids, only 20–30% of the injected Am(III) and Pu(IV) were recovered, whilst with bentonite colloids, 70–85% were recovered; in both cases transport was faster than that of dissolved species. Cs^+ was found to be transported both as a colloidal fraction and in solution, with colloid-mediated transport being more rapid, but Sr^{2+} migration was retarded by sorption to fracture surfaces and was not affected by the presence of colloids.

6 Conclusions

Radioactive contamination in the environment is mainly caused by nuclear weapons production and testing and the nuclear fuel cycle. Historically, emissions to the atmosphere have mainly arisen from weapons testing, causing low-level global contamination from the fallout. Migration in the atmosphere will depend on the nature of the radioactive material and the prevailing meteorological conditions. Within aquatic systems, both terrestrial and surface, a more significant environmental problem is caused by localised high levels of contamination from weapons production and nuclear power. Transport in such environments will be controlled by physical processes such as advection and biogeochemical conditions in the system. In systems with significant flow, advection will be the dominant transport process, but as hydraulic conductivity decreases, chemical processes and conditions become increasingly important in controlling radionuclide migration. Factors such as solution phase chemistry (*e.g.*, ionic strength and ligand concentrations), E_h and the nature of mineral phases in the system have a critical effect on radionuclide speciation, controlling partitioning between solution and solid phases and hence migration. Understanding the complex interplay between these parameters is essential for predicting radionuclide behaviour and migration in the environment.

References

1. J. R. Lloyd and J. C. Renshaw, in *Metal Ions In Biological Systems*, ed. A. Sigel, H. Sigel, and R. K. O. Sigel, Taylor & Francis, Boca Raton, FL, USA, 2005, p. 205.

2. Q.-H. Hu, J.-Q. Weng and J.-S. Wang, *J. Environ. Radioact.*, 2010, **101**, 426.
3. G. R. Choppin, *Radiochim. Acta*, 2003, **91**, 645.
4. UNSCEAR, *Sources and Effects of Ionizing Radiation.* Report of the United Nations Scientific Committee on the Effects of Atomic Radiation to the General Assembly, United Nations, New York, USA, 2008.
5. UNSCEAR, *Report of the United Nations Scientific Committee on the Effects of Atomic Radiation to the General Assembly. Annex C. Exposures to the Public from Man-made Sources of Radiation*, United Nations, New York, USA, 2000.
6. S. L. Simon, A. Bouville and H. L. Beck, *J. Environ. Radioact.*, 2004, **74**, 91.
7. R. C. Ewing, in *Energy, Waste, and the Environment: A Geochemical Perspective*, eds. R. Gieré and P. Stille, The Geological Society, London, UK, 2004, p. 7.
8. G. C. Christensen, G. N. Romanov, P. Strand, B. Salbu, S. V. Malyshev, T. D. Bergan, D. Oughton, E. G. Drozhko, Y. V. Glagolenko, I. Amundsen, A. L. Rudjord, T. O. Bjerk and B. Lind, *Sci. Total Environ.*, 1997, **202**, 237.
9. W. J. F. Standring, D. H. Oughton and B. Salbu, *Environ. Sci. Technol.*, 2002, **36**, 2330.
10. UNSCEAR, *Sources and Effects of Ionizing Radiation - Report to the General Assembly with Scientific Annexes*, United Nations, New York, USA,1993.
11. W. J. F. Standring, O. Stepanets, J. E. Brown, M. Dowdall, A. Borisov and A. Nikitin, *J. Environ. Radioact.*, 2008, **99**, 665.
12. A. F. B. Tompson, C. J. Bruton, G. A. Pawloski, D. K. SMith, W. L. Bourcier, D. E. Shumaker, A. B. Kersting, S. F. Carle and R. M. Maxwell, *Environ. Geol.*, 2002, **42**, 235.
13. National Securities Technologies, *Nevada Test Site Environmental Report Summary*, US Department of Energy, Nevada Test Site, Nevada, USA, 2008.
14. D. K. Smith, D. L. Finnegan and S. M. Bowen, *J. Environ. Radioact.*, 2003, **67**, 35.
15. J. C. Renshaw, J. R. Lloyd and F. R. Livens, *C. R. Chim.*, 2007, **10**, 1067.
16. A. Abdelouas, *Elements*, 2006, **2**, 335.
17. K. Morris and R. Raiswell, in *Interactions of Microorganisms with Radionuclides*, ed. M. J. Keith-Roach and F. R. Livens, Elsevier, London, UK, 2002, 101.
18. H. McKenzie and J. McCord, *Groundwater Annual Report*, Sellafield Ltd, Cumbria, UK, 2009.
19. J. E. Banaszak, B. E. Rittmann and D. T. Reed, *J. Radioanal. Nucl. Chem.*, 1999, **241**, 385.
20. G. A. M. Webb, R. W. Anderson and M. J. S. Gaffney, *J. Radiol. Prot.*, 2006, **26**, 33.
21. S. Van Der Stricht and A. Janssens, *Radioactive Effluents from Nuclear Power Stations and Nuclear Fuel Reprocessing Sites in the European Union, 1999–2003*, European Commission Directorate-General for Energy and Transport, Luxembourg, 2005.

22. J. Gray, S. R. Jones and A. D. Smith, *J. Radiol. Prot.*, 1995, **15**, 99.
23. E. R. Landa, *J. Environ. Radioact.*, 2004, **77**, 1.
24. G. M. Mudd, *J. Environ. Radioact.*, 2008, **99**, 288.
25. S. Handley-Sidhu, M. J. Keith-Roach, J. R. Lloyd and D. J. Vaughan, *Sci. Total Environ.*, 2010, **408**, 5690.
26. A. Bleise, P. R. Danesi and W. Burkart, *J. Environ. Radioact.*, 2003, **64**, 93.
27. UNEP, *Technical Report on Capacity-building for the Assessment of Depleted Uranium in Iraq*, United Nations Environmental Program, Geneva, Switzerland, 2007.
28. W. Briner, *Int. J. Environ. Res. Public Health*, 2010, **7**, 303.
29. K. Rozanski and K. Froehlich, *IAEA Bulletin*, 1996, **9**.
30. IAEA, *Naturally occurring radioactive material (NORM V)*, in *Proceedings of the Fifth International Symposium on Naturally Occurring Radioactive Material*, IAEA, Seville, Spain, 2007.
31. B. Michalik, *J. Environ. Monitor.*, 2009, **11**, 1825.
32. I. Othman and M. S. Al-Masri, *Appl. Radiat. Isot.*, 2007, **65**, 131.
33. M. Betti, L. Aldave de las Heras, A. Janssens, E. Henrich, G. Hunter, M. Gerchikov, M. Dutton, A. W. van Weers, S. Nielsen, J. Simmonds, A. Bexon and T. Sazykina, *J. Environ. Radioact.*, 2004, **74**, 255.
34. C. Papastefanou, *J. Environ. Radioact.*, 2010, **101**, 191.
35. M. Flues, V. Moraes and B. P. Mazzilli, *J. Environ. Radioact.*, 2002, **63**, 285.
36. K. Kant and S. K. Chakarvarti, *Iran J. Radiat. Res.*, 2003, **1**, 133.
37. L. Roberts, *Science*, 1987, **238**, 1028.
38. L. A. Vinhas, in *Security of Radioactive Sources*, IAEA, Vienna, Austria, 2003.
39. National Cancer Institute, *Radon Cancer: Questions and Answers*, 2004, http://www.cancer.gov/cancertopics/factsheet/Risk/radon (accessed 31st October 2010).
40. P. D. Wilson, *The Nuclear Fuel Cycle – From Ore to Waste*, Oxford University Press, Oxford, UK, 1996.
41. J. R. Lloyd and L. E. Macaskie, in *Environmental Microbe–Metal Interactions*, ed. D. R. Lovley, ASM Press, Washington DC, USA, 2000, 277.
42. H. Koch-Steindl and G. Pröhl, *Radiat. Environ. Biophys.*, 2001, **40**, 93.
43. K. O. Konhauser, R. J. G. Mortimer, K. Morris and V. Dunn, in *Interactions of Microorganisms with Radionuclides*, ed. M. J. Keith-Roach and F. R. Livens, Elsevier, Oxford, UK, 2002, p. 61.
44. M. D. Siegel and C. R. Bryan, in *Treatise on Geochemistry Volume 9: Environmental Geochemistry*, ed. B. S. Lollar, Elsevier, Oxford, UK, 2003, p. 205.
45. K. H. Lieser, in *Nuclear and Radiochemistry*, ed. K. H. Lieser, Wiley-VCH, Weinheim Germany, 2001, p. 395.
46. R. N. Collins and A. S. Kinsela, *Chemosphere*, 2010, **79**, 763.
47. S. C. Brooks, D. L. Taylor and P. M. Jardine, *Geochim. Cosmochim. Acta*, 1996, **60**, 1899.
48. B. Salbu, B. Lind and L. Skipperud, *J. Environ. Radioact.*, 2004, **74**, 233.

49. H. L. Beck and B. G. Bennett, *Health Phys.*, 2002, **82**, 591.
50. B. G. Bennett, *Health Phys.*, 2002, **82**, 644.
51. UNEP, *Depleted Uranium in Serbia and Montenegro: Post-conflict Environmental Assessment in the Federal Republic of Yugoslavia*, United Nations Environment Programme, Geneva, Switzerland, 2002.
52. UNEP, *Depleted Uranium in Kosovo: Post-conflict Environmental Assessment*, United Nations Environment Programme, Geneva, Switzerland, 2001.
53. UNEP, *Depleted Uranium in Bosnia & Herzegovina: Post-conflict Environmental Assessment*, United Nations Environment Programme, Geneva, Switzerland, 2003.
54. N. S. Lloyd, S. R. N. Chenery and R. R. Parrish, *Sci. Total Environ.*, 2009, **408**, 397.
55. N. S. Lloyd, J. F. W. Mosselmans, R. R. Parrish, S. R. N. Chenery, S. V. Hainsworth and S. J. Kemp, *Mineral. Mag.*, 2009, **73**, 495.
56. R. R. Parrish, M. Horstwood, J. G. Arnason, S. R. N. Chenery, T. Brewer, N. S. Lloyd and D. O. Carpenter, *Sci. Total Environ.*, 2008, **390**, 58.
57. A. S. Rood, P. G. Voilleque, S. K. Rope, H. A. Grogan and J. E. Till, *J. Environ. Radioact.*, 2008, **99**, 1258.
58. B. G. Lottermoser and P. M. Ashley, *Aust. J. Earth Sci.*, 2006, **53**, 485.
59. UNSCEAR, *Report of the United Nations Scientific Committee on the Effects of Atomic Radiation to the General Assembly. Annex B: Exposures from Natural Radiation Sources*, United Nations, Vienna, Austria, 2000.
60. B. Salbu, *J. Environ. Radioact.*, 2009, **100**, 283.
61. G. E. Brown, A. L. Foster and J. D. Ostergren, *Proc. Natl. Acad. Sci. U. S. A.*, 1999, **96**, 3388.
62. L. A. Warren and E. A. Haack, *Earth Sci. Rev.*, 2001, **54**, 261–320.
63. C. W. Fetter, *Contaminant Hydrogeology*, Prentice Hall International, London, UK, 1998.
64. W. Runde, *Los Alamos Science*, 2000, **26**, 392.
65. K. Morris, J. C. Butterworth and F. R. Livens, *Estuarine Coastal Shelf Sci.*, 2000, **51**, 613.
66. G. A. Waychunas, C. S. Kim and J. F. Banfield, *J. Nanopart. Res.*, 2005, **7**, 409.
67. T. Arnold, S. Utsunomiya, G. Geipel, R. C. Ewing, N. Baumann and V. Brendler, *Environ. Sci. Technol.*, 2006, **40**, 4646.
68. D. M. Singer, K. Maher and G. E. Brown, *Geochim. Cosmochim. Acta*, 2009, **73**, 5989.
69. K. H. Lieser, *Radiochim. Acta*, 1995, **70-1**, 355.
70. S. Langley, A. G. Gault, A. Ibrahim, Y. Takahashi, R. Renaud, D. Fortin, I. D. Clark and F. G. Ferris, *Environ. Sci. Technol.*, 2009, **43**, 1008.
71. N. Sahai, S. A. Carroll, S. Roberts and P. A. O'Day, *J. Colloid Interface Sci.*, 2000, **222**, 198.
72. R. H. Parkman, J. M. Charnock, F. R. Livens and D. J. Vaughan, *Geochim. Cosmochim. Acta*, 1998, **62**, 1481.

73. J. P. McKinley, J. M. Zachara, S. M. Heald, A. Dohnalkova, M. G. Newville and S. R. Sutton, *Environ. Sci. Technol.*, 2004, **38**, 1017.
74. J. K. Fredrickson and J. M. Zachara, *Geobiology*, 2008, **6**, 245.
75. A. de Koning, A. V. Konoplev and R. N. J. Comans, *Appl. Geochem.*, 2007, **22**, 219.
76. B. C. Bostick, M. A. Vairavamurthy, K. G. Karthikeyan and J. Chorover, *Environ. Sci. Technol.*, 2002, **36**, 2670.
77. T. G. Hinton, D. I. Kaplan, A. S. Knox, D. P. Coughlin, R. V. Nascimento, S. I. Watson, D. E. Fletcher and B. J. Koo, *Environ. Sci. Technol.*, 2006, **40**, 4500.
78. E. D. van Hullebusch, P. N. L. Lens and H. H. Tabak, *Rev. Environ. Sci. Bio/Technol.*, 2005, **4**, 185.
79. M. O. Barnett, P. M. Jardine, S. C. Brooks and H. M. Selim, *Soil Sci. Soc. Am. J.*, 2000, **64**, 908.
80. J. P. Bellenger and S. Staunton, *J. Environ. Radioact.*, 2008, **99**, 831.
81. C. Anderson, A. M. Jakobsson and K. Pedersen, *Environ. Sci. Technol.*, 2007, **41**, 830.
82. M. J. Wilkins, F. R. Livens, D. J. Vaughan and J. R. Lloyd, *Biogeochemistry*, 2006, **78**, 125.
83. D. R. Lovley, E. J. P. Phillips, Y. A. Gorby and E. R. Landa, *Nature*, 1991, **350**, 413.
84. J. C. Renshaw, N. Law, A. Geissler, F. R. Livens and J. R. Lloyd, *Biogeochemistry*, 2009, **94**, 191.
85. B. E. Rittmann, J. E. Banaszak and D. T. Reed, *Biodegradation*, 2002, **13**, 329.
86. J. R. Lloyd, P. Yong and L. E. Macaskie, *Environ. Sci. Technol.*, 2000, **34**, 1297.
87. F. R. Livens, M. J. Jones, A. J. Hynes, J. M. Charnock, J. F. W. Mosselmans, C. Hennig, H. Steele, D. Collison, D. J. Vaughan, R. A. D. Pattrick, W. A. Reed, L. N. Moyes, *Proceedings of the International Conference on Radioactivity in the Environment*, Monaco, 2002.
88. R. E. Wildung, S. W. Li, C. J. Murray, K. M. Krupka, Y. Xie, N. J. Hess and E. E. Roden, *FEMS Microbiol. Ecol.*, 2004, **49**, 151.
89. M. J. Wharton, B. Atkins, J. M. Charnock, F. R. Livens, R. A. D. Pattrick and D. Collison, *Appl. Geochem.*, 2000, **15**, 347.
90. K. H. Nealson and D. Saffarini, *Annu. Rev. Microbiol.*, 1994, **48**, 311.
91. J. E. Kostka, E. Haefele, R. Viehweger and J. W. Stucki, *Environ. Sci. Technol.*, 1999, **33**, 3127.
92. C. M. Hansel, S. G. Benner, J. Neiss, A. Dohnalkova, R. K. Kukkadapu and S. Fendorf, *Geochim. Cosmochim. Acta*, 2003, **67**, 2977.
93. K. Hama, K. Bateman, P. Coombs, V. L. Hards, A. E. Milodowski, J. M. West, P. D. Wetton, H. Yoshida and K. Aoki, *Clay Minerals*, 2001, **36**, 599.
94. I. T. Burke, F. R. Livens, J. R. Lloyd, A. P. Brown, G. T. W. Law, J. M. McBeth, B. L. Ellis, R. S. Lawson and K. Morris, *Appl. Geochem.*, 2010, **25**, 233.

95. M. J. Wilkins, F. R. Livens, D. J. Vaughan, I. Beadle and J. R. Lloyd, *Geobiology*, 2007, **5**, 293.
96. I. T. Burke, C. Boothman, J. R. Lloyd, F. R. Livens, J. M. Charnock, J. M. McBeth, R. J. G. Mortimer and K. Morris, *Environ. Sci. Technol.*, 2006, **40**, 3529.
97. K. Morris, F. R. Livens, J. M. Charnock, I. T. Burke, J. M. McBeth, J. D. C. Begg, C. Boothman and J. R. Lloyd, *Appl. Geochem.*, 2008, **23**, 603.
98. T. Borch, R. Kretzschmar, A. Kappler, P. Van Cappellen, M. Ginder-Vogel, A. Voegelin and K. Campbell, *Environ. Sci. Technol.*, 2010, **44**, 15.
99. H. Dong, R. K. Kukkadapu, J. K. Fredrickson, J. M. Zachara, D. W. Kennedy and H. M. Kostandarithes, *Environ. Sci. Technol.*, 2003, **37**, 1268.
100. R. S. Cutting, V. S. Coker, J. W. Fellowes, J. R. Lloyd and D. J. Vaughan, *Geochim. Cosmochim. Acta*, 2009, **73**, 4004.
101. J. Hopf, F. Langenhorst, K. Pollok, D. Merten and E. Kothe, *Chem. Erde-Geochem.*, 2009, **69**, 45.
102. J. E. Kostka, D. D. Dalton, H. Skelton, S. Dolhopf and J. W. Stucki, *Appl. Geochem.*, 2002, **68**, 6256.
103. J. Kim, H. L. Dong, J. Seabaugh, S. W. Newell and D. D. Eberl, *Science*, 2004, **303**, 830.
104. D. P. Jaisi, H. L. Dong and C. X. Liu, *Geochim. Cosmochim. Acta*, 2007, **71**, 1145.
105. S. Langley, A. G. Gault, A. Ibrahim, Y. Takahashi, R. Renaud, D. Fortin, I. D. Clark and F. G. Ferris, *Chem. Geol.*, 2009, **262**, 217.
106. D. Fortin and S. Langley, *Earth-Sci. Rev.*, 2005, **72**, 1.
107. B. M. Tebo, J. R. Bargar, B. G. Clement, G. J. Dick, K. J. Murray, D. Parker, R. Verity and S. M. Webb, *Annu. Rev. Earth Planet. Sci.*, 2004, **32**, 287.
108. S. M. Webb, C. C. Fuller, B. M. Tebo and J. R. Bargar, *Environ. Sci. Technol.*, 2006, **40**, 771.
109. R. J. Silva and H. Nitsche, *Radiochim. Acta*, 1995, **70–71**, 377.
110. G. R. Choppin, *Marine Chem.*, 2006, **99**, 83.
111. C. C. Choy, G. P. Korfaitis and X. Meng, *J. Hazard. Mater.*, 2006, **136**, 53.
112. D. G. Brookins, *Eh-pH Diagrams for Geochemistry*, Springer-Verlag, Berlin, Germany, 1988.
113. C. E. W. Steinberg, S. Kamara, V. Y. Prokshotskaya, L. Manusadzianas, T. A. Karasyova, M. A. Timofeyev, Z. Jie, A. Paul, T. Meinelt, V. F. Farjalla, A. Y. O. Matsuo, B. K. Burnison and R. Menzel, *Freshwater Biol.*, 2006, **51**, 1189.
114. R. Sutton and G. Sposito, *Environ. Sci. Technol.*, 2005, **39**, 9009.
115. P. E. Reiller, N. D. M. Szabo and G. Szabo, *Radiochim. Acta*, 2008, **96**, 345.
116. R. E. Keepax, D. M. Jones, S. E. Pepper and N. D. Bryan, in *Interactions of Microorganisms with Radionuclides*, ed. M. J. Keith-Roach and F. R. Livens, Elsevier, London, UK, 2002, p. 143.
117. K. Nash, S. Fried, A. M. Friedman and J. C. Sullivan, *Environ. Sci. Technol.*, 1981, **15**, 834.
118. V. Moulin and C. Moulin, *Appl. Geochem.*, 1995, **10**, 573.

119. S. Uroz, C. Calvaruso, M. P. Turpault and P. Frey-Klett, *Trends Microbiol.*, 2009, **17**, 378.
120. J. M. Arocena, L. P. Zhu and K. Hall, *Earth Surf. Processes Landforms*, 2003, **28**, 1429.
121. M. J. Keith-Roach, *Sci. Total Environ.*, 2008, **396**, 1.
122. W. D. Samuels, D. M. Camaioni and H. Babad, *Initial Laboratory Studies into the Chemical and Radiological Aging of Organic Materials in Underground Storage Tanks at the Hanford Complex*, Pacific Northwest National Laboratory, Richland Washington, USA, 1998.
123. H. Boukhalfa, G. A. Icopini, S. D. Reilly and M. P. Neu, *Appl. Environ. Microbiol.*, 2007, **73**, 5897.
124. P. A. Rusin, L. Quintana, J. R. Brainard, B. A. Strietelmeier, C. D. Tait, S. A. Ekberg, P. D. Palmer, T. W. Newton and D. L. Clark, *Environ. Sci. Technol.*, 1994, **28**, 1686.
125. Y. A. Gorby, J. F. Caccavo and H. Bolton, *Environ. Sci. Technol.*, 1998, **32**, 244.
126. J. F. Caccavo, D. J. Lonergan, D. R. Lovley, M. Davis, J. F. Stolz and M. J. McInerney, *Appl. Environ. Microbiol.*, 1994, **60**, 3752.
127. I. AlMahamid, K. A. Becraft, N. L. Hakem, R. C. Gatti and H. Nitsche, *Radiochim. Acta*, 1996, **74**, 129.
128. B. A. Logue, R. W. Smith and J. C. Westall, *Environ. Sci. Technol.*, 2004, **38**, 3752.
129. G. R. Choppin, *Czech. J. Phys.*, 2006, **56**(Suppl. D), D13.
130. F. Heberling, M. A. Denecke and D. Bosbach, *Environ. Sci. Technol.*, 2008, **42**, 471.
131. S. D. Kelly, M. G. Newville, L. Cheng, K. M. Kemner, S. R. Sutton, P. Fenter, N. C. Sturchio and C. Spotl, *Environ. Sci. Technol.*, 2003, **37**, 1284.
132. S. D. Kelly, E. T. Rasbury, S. Chattopadhyay, A. J. Kropf and K. M. Kemner, *Environ. Sci. Technol.*, 2006, **40**, 2262.
133. J. Friedrich, *J. Marine Systems*, 2009, **76**(special issue (SI), 322.
134. M. C. Duff, J. U. Coughlin and D. B. Hunter, *Geochim. Cosmochim. Acta*, 2002, **66**, 3533.
135. R. Kretzshmar and T. Schäfer, *Elements*, 2005, **1**, 205.
136. F. Eyrolle and S. Charmasson, *J. Environ. Radioact.*, 2001, **55**, 145.
137. H. Geckeis, B. Grambow, A. Loida, B. Luckscheiter, E. Smailos and J. Quinones, *Radiochim. Acta*, 1998, **82**, 123.
138. J. F. McCarthy and L. D. McKay, *Vadose Zone J.*, 2004, **3**, 326.
139. J. N. Ryan and M. Elimelech, *Colloids Surf., A*, 1996, **107**, 1.
140. H. Geckeis and T. Rabung, *J. Contam. Hydrol.*, 2008, **102**, 187.
141. A. P. Novikov, S. P. Kalmykov, S. Utsunomiya, R. C. Ewing, F. Horread, A. Merkulov, S. B. Clark, V. V. Tkachev and B. F. Myasoedov, *Science*, 2006, **314**, 638.
142. A. Möri, W. R. Alexander, H. Geckeis, W. Hauser, T. Schäfer, J. Eikenberg, T. Fierz, C. Degueldre and T. Missana, *Colloids Surf., A*, 2003, **217**, 33.
143. Q.-H. Hu, M. Zavarin and T. P. Rose, *Geochem. Trans.*, 2008, **9**, 1.

Radiation Protection of the Environment: A Summary of Current Approaches for Assessment of Radionuclides in Terrestrial Ecosystems

B. J. HOWARD* AND N. A. BERESFORD

ABSTRACT

Over the past decade the international community has recognised the need to demonstrate that wildlife populations are protected from environmental releases of radioactivity as well as humans. Frameworks and models for such assessments have been developed and are continuously being tested and improved. In this chapter, the basic elements of an assessment for radiation exposure of wildlife are outlined, including the current methods used to estimate environmental radionuclide transfer and the resulting doses. The methods used to derive benchmarks based on radiation effects data, against which estimated doses can be compared, are described. Since it is impossible to quantify transfer and doses for all species, the approaches use representative groups such as "reference organisms" including the Reference Animals and Plants of the International Commission on Radiological Protection (ICRP). The current approaches used for wildlife have some commonalities with those used for humans, but with some notable differences. Organisms tend to be considered as homogenous, simplified geometric shapes with the whole organism absorbed dose rate being estimated; the majority of available effects data are expressed on the basis of whole organism dose rates. Transfer is often quantified by predicting the whole organism activity concentration from that in the environmental media such as soil, water or air. Protection is focused on

* Corresponding author

Issues in Environmental Science and Technology, 32
Nuclear Power and the Environment
Edited by R.E. Hester and R.M. Harrison

Published by the Royal Society of Chemistry, www.rsc.org

populations rather than individuals, and therefore some approaches used for the assessment of chemicals pollutants are also being adopted for radionuclides.

1 Introduction

Man-made radionuclides are emitted from the nuclear fuel cycle into the atmosphere in gaseous discharges and into aquatic ecosystems *via* liquid discharges. Radionuclides can be transferred from air, soil, water and sediment to organisms. Liquid discharges input radionuclides into water bodies from which they can be transferred to sediment, which is the major sink in aquatic systems. Gaseous discharges are deposited onto both plant and soil surfaces, and soil is the major reservoir for most radionuclides in the terrestrial environment. To illustrate various issues in the chapter, we will concentrate on a few radionuclides: ^{60}Co, ^{90}Sr, ^{137}Cs, $^{239+240}$Pu, ^{241}Am and ^{131}I. Noble gases, which are discharged in comparatively high amounts, are not important in dose terms, due to their low environmental transfer and dose coefficients (see the predicted dose rates presented by Copplestone *et al.*).[1]

In this chapter, we describe the major terrestrial pathways for radionuclides released during the nuclear fuel cycle, with a focus on wildlife rather than the human food chain. We briefly describe the approaches being used and developed to demonstrate protection of wildlife from releases of radioactivity into the environment. Key differences in the approaches to predicting environmental transfer between the human food chain and environmental assessment methods for wildlife will be highlighted. More detailed information on this topic can be accessed at http://www.ceh.ac.uk/protect.

The major factors affecting the extent of transfer of radionuclides to organisms are briefly described. In many recent radiological documents, the term "non-human" biota has been used for organisms other than humans. This term is rarely used in ecotoxicology and other areas of environmental protection. Here we use the term "wildlife" which encompasses wild plants, undomesticated animals and organisms such as fungi and bacteria (*i.e.* the potential objects of environmental protection).

The degree of internal exposure arising from man-made radionuclides in the environment depends on the environmental behaviour of the radionuclides emitted. The environmental mobility of different radionuclides varies considerably. Radionuclides with a potentially high environmental mobility include ^{131}I, $^{134/137}$Cs, ^{90}Sr, ^{14}C, ^{3}H, ^{35}S, whereas those with low environmental mobility include $^{239/240}$Pu and ^{241}Am. Many different factors affect environmental mobility and the extent to which radionuclides are transferred into organisms. We briefly describe a range of these processes but as an example focus on those which are most important in determining transfer to wildlife.

2 Radiation Protection of Wildlife

For many years, protection of the environment from radiation was anthropocentric based on the ICRP statement[2] that

"The Commission believes that the standard of environmental control needed to protect man to the degree currently thought desirable will ensure that other species are not put at risk. Occasionally, individual members of non-human species might be harmed, but not to the extent of endangering whole species or creating imbalance between species. At the present time, the Commission concerns itself with mankind's environment only with regard to the transfer of radionuclides through the environment, since this directly affects the radiological protection of man".

Thus, the protection criterion for humans ($1\,\mathrm{mSv\,y^{-1}}$; see the following chapter by Pentreath) was considered to be sufficiently restrictive that populations of non-humans living in the same environment would be sufficiently protected.

Over the last decade, systems of radiological protection for wildlife have begun to evolve with considerable international and national effort on this issue. In the 2007 Recommendations of the ICRP, the Commission recommended the explicit consideration of Radiological Protection of the Environment and recognised the need for advice and guidance, including a clearer framework.[3] In 2005, the ICRP formed a fifth committee, which deals specifically with the protection of the environment from ionising radiation. Committee 5 proposed a framework for protection of the environment which uses the concept of Reference Animals and Plants (RAPs), designed to be compatible with the system of protection used for humans.[4] The ICRP also aims to produce a system similar to those used for protection of the environment from other hazards.

The need for a system capable of demonstrating that the environment is adequately protected from the effects of radioactive substances has been recognised by international organisations (*e.g.* the International Atomic Energy Agency; IAEA)[5] and a number of regulators. Environmental protection is now referred to in the International Atomic Energy Agency's Fundamental Safety Principles.[6] The forthcoming revision of the International Basic Safety Standards also mentions radiological protection of the environment. Different approaches have been developed to estimate the exposure of wildlife to ionising radiation. These approaches are used, in some countries, to address requirements in national legislation to demonstrate that the environment is protected from anthropogenic releases of radioactivity.[7–12] Radiation protection has not always been the driver of this process, in some countries a system of protection is required to address conservation legislation.

The ICRP[4] has focussed on Reference Animals and Plants (RAPs) which are defined as follows

"A hypothetical entity, with the assumed basic biological characteristics of a particular type of animal or plant, as described to the generality of the taxonomic level of Family, with defined anatomical, physiological, and life-history properties, that can be used for the purposes of relating exposure to dose, and dose to effects, for that type of living organism".

The RAPs are hypothetical entities and not intended to represent a particular species. Commonly, in assessment approaches linked to tools, the approach taken to address the wide range of different wildlife species is to use "reference organisms". The selection of reference organisms has considered the need to encompass protected species, and different trophic levels and exposure pathways.[13,14] Reference organisms have tended to be defined at a broad wildlife group level (*e.g.* soil invertebrate, predatory fish, terrestrial mammal *etc.*). The definition of reference organisms used in the integrated Environmental Risk from Ionizing Contaminants Assessment and Management (ERICA) approach developed by EC researchers[15] is

> *"a series of entities that provide a basis for the estimation of radiation dose rate to a range of organisms which are typical, or representative, of a contaminated environment. These estimates, in turn, would provide a basis for assessing the likelihood and degree of radiation effects".*

In contrast, for some approaches specific species rather than wide groupings have been considered.[11,16]

To assess the risk of radioactivity to wildlife we need an approach which contains the following components:

(i) transfer of radionuclides to wildlife;
(ii) dose conversion coefficients relating internal and media activity concentrations to estimate absorbed dose rates to wildlife; and
(iii) interpretation of the biological effects of radiation to determine the risk to wildlife.

There are currently three comparatively comprehensive assessment models which are freely available: the ERICA Tool which implements the ERICA Integrated Approach,[17] RESRAD-BIOTA[18] which implements the US Department of Energy's Graded Approach[9] and the England and Wales Environment Agencies R&D128.[11,14]

3 Environmental Transfer in Terrestrial Ecosystems

To be able to estimate radionuclide activity concentrations in exposed organisms we need to quantify and model the transfer processes. The approach used varies depending on the objective and the need for detailed information. Some models mathematically describe the transfer processes through the application of steady state compartment models which assume that there is an equilibrium established in the environment between the source and the receptor. However, in some cases there may be a need to describe transfer in more detail, either to take account of some of the many environmental factors which affect the extent of transfer in the environment, or to quantify changes in transfer with time after radionuclides have been received by ecosystems.

The pathways leading to exposure of organisms in terrestrial ecosystems can be subdivided into external (see section 4) and internal components. Internal irradiation occurs from radionuclides which are absorbed and distributed within the organism. Ingestion of plants, animals, soil/sediment and detritus also leads to direct irradiation of the digestive tract.

Since the dosimetric calculation to estimate absorbed doses are derived for a defined shape of a whole organism (see section 4), there is a requirement to estimate the whole organism activity concentrations. This contrasts with the human food chain where the focus is on the part of the organism that is ingested by humans.

The processes involved in environmental terrestrial pathways are briefly summarised here before we describe the current methods by which exposure of organisms to radionuclides are currently quantified and evaluated.

3.1 Atmospheric Deposition

Following atmospheric release, vegetation intercepts radionuclides from wet, dry or occult deposition[19] and the remaining radionuclides are deposited to the ground surface. The fraction of radionuclides intercepted by vegetation is dependent on the developmental stage of the plant and the amount of above ground biomass, and consequently the time of year is important in determining how much radionuclide is retained initially on plant surfaces. Leafy vegetables, because of their large surface area, have a high interception of radionuclides which is currently being demonstrated around the Fukushima site where radioiodine and radiocaesium activity concentrations in spinach are high compared with other crops.[66]

For dry deposition, interception is more effective for small particles and reactive gases than for larger particles. Interception of wet-deposited radionuclides is a result of the complex interaction of the chemical form of the element and the stage of development of the plant.

The process of loss of radionuclides from plant surfaces is termed "weathering", which is influenced by a number of physical processes, including wash-off by rain (or irrigation in agricultural systems), surface abrasion, wind action, tissue senescence, leaf fall, herbivore grazing, growth, volatilisation and evaporation.[20]

Direct ingestion by animals of radionuclides intercepted by vegetation can be an important contributor to radionuclide intake.

3.2 Radionuclides in Soil

The mobility of most radionuclides from soil to other organisms is predominantly *via* plant root uptake, which will be largely determined by physicochemical factors influencing the distribution of radionuclides between the solid and solution phases of soil. The uptake of most elements by plant roots occurs mainly from the soil solution.

The important interactions of any chemical species in solution, which can influence its mobility in soils and eventual root uptake, include: charge interactions; complexation and precipitation reactions with other chemical species (*e.g.* organic and inorganic ligands); oxidation–reduction (redox) transformations; and specific interactions with soil components including soil biota. Soil factors influencing the mobility of some of these radionuclides are outlined in the following paragraph.

The extent of sorption in soil is described by the solid–liquid distribution coefficient (K_d).

$$K_d\ (\mathrm{l\,kg^{-1}}) = \frac{\text{Activity concentration in soil }(\mathrm{Bq\,kg^{-1}}\text{ dry weight})}{\text{Activity concentration in filtered water }(\mathrm{Bq\,l^{-1}})} \qquad (1)$$

Simple K_d-based models assume that the radionuclide on the solid phase is in equilibrium with that in solution. However, K_d can change with time as the sorption process "ages".

The K_d for a radionuclide may vary within various orders of magnitude depending on the combination of radionuclide and soil type.[21] The use of a cofactor approach can decrease the variability of the ranges of K_d values associated with a soil type. For example, K_d is affected by the radiocaesium interception potential (RIP), K and NH_4^+ status for radiocaesium, the cation exchange capacity (CEC), Ca and Mg concentrations for radiostrontium, and the pH for heavy radionuclides.[21]

Caesium is strongly sorbed in soil by ion exchange, some of which is irreversible, or fixed, with fixation being influenced by clay mineralogy. A number of models relating the availability of radiocaesium to soil properties have been proposed, including increasing soil–plant transfer with increasing soil organic matter;[22,23] decreasing soil–plant transfer with increasing soil solution potassium;[24] a semi-mechanistic approach using soil clay and organic matter contents, exchangeable K status, pH and NH_4^+ concentration;[25] and, more recently, the use of RIP of soils and exchangeable potassium concentration to predict caesium uptake.[21] Characterisation of the soils in the areas of Japan which are receiving radionuclide deposition from Fukushima should enable reasonable predictions of the long term availability of radiocaesium to foodstuffs.

Recent comparisons of data[21] showed that the K_d of strontium for sand, loam, clay and organic soil groups were similar, although the value for the sand group was significantly lower. For radiostrontium, the key soil characteristics determining sorption were CEC and calcium and magnesium concentrations.

The transuranic radionuclides, americium and plutonium, have relatively low mobility due to their strong tendency to sorb onto soil particles.[26] Americium generally exists in the III and/or IV valence state, whereas plutonium often exists in the IV state, but can be found in any of four oxidation states (III, IV, V or VI) depending on the redox conditions of the soil system.

3.3 Radionuclide Transfer to Plants

Radionuclides can be transferred to plants *via* stomatal uptake and cuticular absorption, often referred to as "foliar uptake". For many radionuclides, however, "root uptake" is more important than foliar uptake, although the latter can be significant for some radionuclides if there is a continuous aerial discharge. After foliar or root uptake, radionuclides are translocated *via* the phloem to different parts of plants.

The processes in soil described above lead to variation in plant uptake due to the differing proportions of radionuclides available in the soil solution. For example, the transfer of radiocaesium from soil to many plants follows the order: clay < loam < sand < organic soils, but there is considerable variation within, and overlap between, the four soil groups. Plant uptake of americium is generally ten-fold higher than that of plutonium.[26]

The uptake of radionuclides by plants (fungi and microbes) also varies between different species, as well as with the soil type.

3.3.1 Quantification of Transfer to Plants

Radionuclide transfer to plants in the human food chain is often quantified using concentration ratios for different groups of plants and soil types. A recent IAEA handbook provides compiled concentration ratio values for the human food chain.[27,28] There is an assumption that equilibrium exists between the plant and soil, which is not valid if there are large temporal spikes in releases.

Similarly, for estimation of exposure of plants themselves in environmental assessment models, the most common approach is the whole organism concentration ratio (CR_{wo}), where:

$$CR_{wo} = \frac{\text{Activity concentration in whole organism (Bq kg}^{-1}\text{ fresh weight)}}{\text{Activity concentration in soil (Bq kg}^{-1}\text{ dry weight)}} \qquad (2)$$

In human food chain models, CR is most usually defined on the basis of plant dry matter activity concentrations. The CR values, categorised by soil type, for human food chain assessments have been collated for many radionuclides in the IAEA handbook.[28] More mechanistically based approaches enabling predictions of radionuclides, such as radiocaesium and radiostrontium, which vary with soil properties are also available in some assessment tools for the human food chain. In contrast, current CR values for estimation of plant exposure generally do not distinguish between different soil types.

Whilst root uptake is a key pathway of plant contamination, for radionuclides which have a low root uptake, such as plutonium and americium, resuspension and adherence of contaminated soil on plant surfaces can constitute a significant proportion of the radionuclide content of plants as sampled in the environment.

As examples, the CR values for the selected radionuclide–wildlife group combinations as anticipated to be reported in the new IAEA handbook on transfer to wildlife are given in Figure 1.

3.4 Radionuclide Transfer to Terrestrial Animals

Animals can be contaminated through the skin, by inhalation, and, most importantly, *via* ingestion of radionuclides. Uptake through the skin is not usually an important route of contamination, and is not considered here. Inhalation by animals is potentially more important than skin absorption since the lung surfaces, the site of gaseous exchange, are more permeable to a wider range of elements. Radionuclides may be inhaled in different forms, including gaseous compounds, aerosols and particles by terrestrial animals. The ability of radionuclides to pass through the pulmonary membranes varies considerably. Despite low transfer rates for actinides, such as plutonium, they are often more readily absorbed *via* the lungs than the gastrointestinal tract. Gaseous iodine is readily absorbed and inhalation may have been a route of contamination of milk of housed animals following the Chernobyl accident and currently in Japan.[67] Inhalation, however, is generally not a major contamination route for most radionuclides for animals and is not considered further here.

The most important transfer pathway to animals is the ingestion of contaminated food, soil and drinking water. Intake *via* drinking water is generally a small contributor to total radionuclide intake. Radionuclide intake *via* soil can be significant, but the availability for absorption of soil-associated radionuclides may be lower than plant incorporated radionuclides, although there is only evidence of such a difference for caesium.[29] Hence, it is the ingestion of contaminated feed and processes influencing absorption and retention that usually determines the radionuclide content of animals.

3.4.1 Gastrointestinal Absorption

The degree of absorption from the gastrointestinal tract is an important factor in determining the radionuclide content of animal tissues.

For mammals, fractional absorption has been quantified as the true absorption coefficient, A_t, which can be determined as the difference in dietary intake and faecal output (corrected for the endogenous secretion of radionuclides into the gastrointestinal tract),[29] expressed as a proportion of dietary intake. Absorption of most essential elements is controlled by dietary supply and the animal's requirement (absorption tending to decrease with increasing dietary concentrations when requirement is met) and, in some instances, other essential elements such as interactions between calcium and phosphorous.

Fractional absorption values for both monogastric mammals and ruminants can be found in papers by the ICRP[30] and Howard *et al.*[31] The absorption of essential elements tends to be relatively high. In contrast, elements with high atomic weights which are not essential elements or analogues of essential elements are generally poorly absorbed.

Most forms of iodine are rapidly reduced to iodide within the digestive tract and absorption is complete regardless of the source of radioiodine or amount of dietary stable iodine.

In the case of radiocaesium, the source ingested is a major factor determining subsequent concentrations in tissues, with the true absorption coefficient ranging from <0.10 to >0.80.[29] Absorption of particle or soil-associated radiocaesium is considerably lower than that of radiocaesium incorporated within plants.

The variation in A_t for radiostrontium, with recorded values ranging from 0.05–0.7, does not appear to be related to source but is strongly influenced by its analogue, calcium, which is a homeostatically controlled essential element. The extent of calcium absorption in the gastrointestinal tract is governed by the animal's calcium requirement which depends on factors such as age, growth rate and milk yield. At a given calcium requirement, there is an inverse relationship between the absorption of calcium and the amount of calcium in the diet.[29] Under normal levels of calcium intake, the source of radiostrontium ingested is unlikely to influence either the extent of absorption or the concentrations in tissues.

The bioavailability of transuranic elements, such as plutonium and americium, is low compared with that of many elements. Fractional absorption values are generally less than 0.0001.

Gastrointestinal fractional absorption decreases with age which may be due to the lower permeability of the membranes of the gastrointestinal wall of mature animals compared with young animals (most notably from birth to a few weeks old) which have a greater need to absorb a wide range of nutrients and essential elements.

3.4.2 Radionuclide Distribution in Animals

Once absorbed, radionuclides enter the circulatory system and are distributed into various tissues of the body. In some cases, radionuclides are biotransformed within tissues and may be present within the animal in more than one form. For instance, ^{3}H occurs as tissue water or organically bound tritium incorporated into the protein and fat of tissues.

Different radionuclides are accumulated in different tissues. For some radionuclides, the site of deposition is determined by the biological role of the corresponding stable element or analogue. The major iodine storage organ in the body is the thyroid and the element is also actively taken up by the mammary gland and transferred into milk. Radiostrontium behaves as a calcium analogue and is therefore accumulated in bone and shell and is also transferred into milk. Radiocaesium is an analogue of potassium and is, therefore, found in all soft tissues. The actinides and rare earth elements are all accumulated in bone. Liver (or hepatopancreas in arthropods and gastropods) and, to a lesser extent, kidneys are common storage tissues for many pollutants including some radionuclides (*e.g.* actinide elements and heavy metals).

3.4.3 Quantification of Transfer to Animals

For human food chains, the transfer of radionuclides to milk and meat has previously been commonly quantified using the transfer coefficient defined as the equilibrium ratio between the radionuclide activity concentration in milk (F_m; $d\,l^{-1}$ or $d\,kg^{-1}$) or meat (F_f; $d\,kg^{-1}$) and the daily dietary radionuclide intake. Transfer coefficients for smaller animals are higher than those for larger animals, and those for adults are lower than those for (smaller) young livestock. Beresford *et al.*[32] suggested that much of this difference observed in transfer coefficients arises because they incorporate dry matter intake which increases with animal size, and suggested that the concentration ratio between the activity concentration in an animal product and diet may be a less variable and more generic parameter (later substantiated in the IAEA handbook[28]). Mean concentration ratios reported in IAEA[28] for milk are highest for Cs (0.15) and for essential elements including I (0.46).[68,69] The transfer of caesium and/or iodine isotopes to milk has required mitigation after the Windscale, Chernobyl and Fukushima accidents.

For wildlife assessment, most approaches use concentration ratios for at least some organisms. The *CR* is based on the whole organism activity concentration for terrestrial animals most usually compared to the soil activity concentration, see Equation (2).

The "whole organism" generally excludes the outer parts such as the skin and feathers, and the gut which can contain ingested material which is more highly contaminated than other body tissues.

Ratios approaches assume equilibrium, but there can be considerable temporal variation in an animal's intake of radionuclides and hence tissue concentrations may be constantly changing. Equilibrium will often not have been reached within an animal's lifetime, especially for radionuclides with long physical and biological half-lives in tissues (*e.g.* plutonium). Dynamic models describing the behaviour of radionuclides within animal tissues have been developed for human food chains which can be used to predict radionuclide activity concentrations in different tissues following continuous, single or varying intakes.[33–36]

Differences in the quantification of transfer in the currently available assessment tools have resulted in large variation in predicted whole organism activity concentrations and resultant internal doses.[37–40] In response, CR_{wo} values have been collated for terrestrial, freshwater, marine and estuarine ecosystems in an online database and the data reported in broad wildlife groups by the IAEA in a Technical Report Series Handbook currently in preparation.[41] Since much of the reported data for activity concentration in organisms are for edible fractions used in the human food chain, the handbook also provides tables to enable the conversion of data for edible fractions to whole organism values.[42]

In the IAEA handbook, all the *CR* values are based on reported data. As an example of the CR_{wo} data available, the values for the selected radionuclides for some terrestrial wildlife groups are shown in Figure 1.

Given the large number of potential radionuclide–organism combinations which may require consideration within an environmental assessment many CR_{wo} values cannot be derived from the literature. Various methods have been proposed to extrapolate from the available data to provide values for missing combinations, such as described by Beresford *et al.*[43] and Higley.[44]

4 Dosimetry for Wildlife

Radionuclides in the environment lead to plants and animals being exposed both externally and internally to ionising radiation. Internal exposure arises from radionuclides incorporated into the organism by the processes described

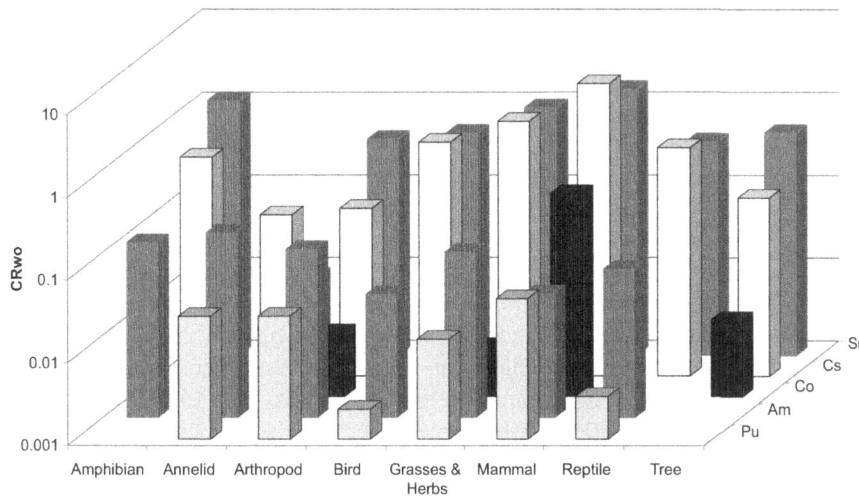

Figure 1 Example CR_{wo} values for different radionuclide–wildlife group combinations (data reproduced from draft IAEA TRS).[41] Am: reptile – carnivorous only, Co: mammal – omnivorous only, Pu: grasses only, no herbs.

in the previous section. In addition to the activity concentration in an organism, internal exposure depends upon the organism size and the type and energy of emitted radiation. External exposure is largely determined by the contamination levels in the environment, habitat (the geometric relationship between the radiation source and the organism and the shielding properties of the medium), organism size, and the physical properties of the radionuclides.

The interaction of radiation with matter leads to the excitation and ionisation of the target material (tissue). The unit of absorbed dose is the Gray (Gy), where one Gy = one Joule of absorbed energy per kg material ($J\,kg^{-1}$). Dose conversion coefficients (DCCs), defined as absorbed dose rate ($\mu Gy\,h^{-1}$) per unit activity concentration in an organism ($Bq\,kg^{-1}$ fw; where fw = fresh weight) or medium (Bq per unit media fw), are used to relate organism and medium activity concentrations of an absorbed dose.

In the simplest case, an organism is assumed to be in an infinite homogeneous medium with the same density as itself, with the radionuclide distributed homogenously throughout all its tissues. Under these conditions, both internal (DCC_{int}) and external (DCC_{ext}) dose conversion coefficients for mono-energetic radiation can be expressed as a function of the absorbed fraction, as follows:

$$DCC_{int} = E \times \phi(E) \tag{3}$$

$$DCC_{ext} = E \times (1 - \phi E) \tag{4}$$

Where E (eV) is the energy of a mono-energetic source and $\phi(E)$ is the absorbed fraction for the energy E.

The equations assume that the organism and the surrounding medium are of similar density and elemental composition. If the radiation is not mono-energetic, the above definition can be generalised by summing the terms over the different radionuclide decay energies, weighted by the branching ratios of each transition. For external exposure, if the organism receives contributions from various environmental media (which may not always be assumed to have the same density as the organism), the equation also needs to sum these individual contributions.

The key quantity for estimating internal absorbed doses is the absorbed fraction (ϕ), defined as the fraction of energy emitted by a radiation source that is absorbed by an organism. The uncertainty associated with the heterogeneous distribution of some radionuclides in organisms has been assessed.[45] The conclusions were that: (i) for photons, the uncertainty due to a possible non-homogeneous radionuclide distribution is lower than 20–25% in the considered cases; and (ii) for electrons, uncertainty is below 30% and likely to be negligible below an energy of 0.5 MeV.

4.1 Dose Rate Calculation

The dose conversion coefficients can be used to estimate the unweighted absorbed dose rate from media and organism activity concentrations. For internal exposure the following equation can be used:

$$\dot{D}^b_{\text{int}} = \sum_i C^b_i * DCC^b_{\text{int},i} \tag{5}$$

Where:

- $\dot{D}^b_{\text{int}}$ is the absorbed internal dose rate for reference organism b;
- C^b_i is the average concentration of radionuclide i in reference organism b (Bq kg^{-1} fw); and
- $DCC^b_{\text{int},i}$ is the radionuclide-specific dose conversion coefficient for internal exposure defined as the ratio between the average activity concentration of radionuclide i in the organism b and the dose rate to the organism (μGy h^{-1} per Bq kg^{-1} fw).

For external exposure the following equation can be used in terrestrial ecosystems:

$$\dot{D}^b_{\text{ext}} = \sum_z v_z \sum_i C^{ref}_{zi} * DCC^b_{\text{ext},zi} \tag{6}$$

Where:

- v_z is the occupancy factor, the fraction of time that organism b spends at a specified location z in its habitat;
- C^{ref}_{zi} is the average concentration of radionuclide i in the reference media of a given location z (Bq kg^{-1} fw, soil); and

- $DCC^{b}_{\text{ext},zi}$ is the dose conversion factor for external exposure defined as the ratio between the average activity concentration of radionuclide i in the reference media corresponding to the location z and the dose rate to organism b ($\mu\text{Gy}\,\text{h}^{-1}$ per Bq unit media).

External DCCs from beta and alpha emitters are comparatively low and may be assumed to be zero for some radionuclides in some approaches.[46]

Weighted total dose rates (in $\mu\text{Gy}\,\text{h}^{-1}$) can be calculated as:

$$DCC_{\text{int}} = wf_{\text{low}\beta} \cdot DCC_{\text{int,low}\beta} + wf_{\beta+\gamma} \cdot DCC_{\text{int},\beta+\gamma} + wf_{\alpha} \cdot DCC_{\text{int},\alpha} \tag{7}$$

$$DCC_{\text{ext}} = wf_{\text{low}\beta} \cdot DCC_{\text{ext,low}\beta} + wf_{\beta+\gamma} \cdot \text{DCC}_{\text{ext},\beta+\gamma} \tag{8}$$

Where wf are the weighting factors for various components of radiation (low energy β, $\beta+\gamma$ and α).

Although there is no agreement on wf for wildlife, currently most assessment approaches are using broadly similar values with default radiation weighting factors of 10–20 for alpha radiation and 1–3 for low beta radiation and 1 for beta-gamma radiation.[13,14,47] For α-radiation weighting factors, these values are broadly consistent with the upper bound on the range of variation reported by Chambers *et al.*[48] in relation to deterministic endpoints (mainly mortality). Currently, the estimated doses for wildlife do not take account of tissue weighting factors as used in human dosimetry.

The ICRP have derived dose conversion coefficients[4] for 75 radionuclides for the Reference Animals and Plants using the methodology of Ulanovsky and Pröhl[49] developed for the ERICA Tool.[13]

Vives i Batlle *et al.*[50] presents a comparison of unweighted whole-body dose rates from ten different approaches being used (or developed) to assess the exposure of wildlife to radiation.

5 Effects on Wildlife

DNA is the primary target for the induction of biological effects from radiation in all living organisms. There are broad similarities in radiation responses from different organisms, but differences in radiation sensitivity. The range in lethality from acute exposure to radiation varies by three to four orders of magnitude amongst organisms, with mammals being among the most sensitive and viruses being among the most radioresistant.[51]

Damage from radiation is initiated by ionisation which occurs if the radiation has sufficient energy to eject one or more orbital electrons from the atom in which it interacts. Ionising radiation is characterized by a large release of energy which can break strong chemical bonds. The ionisation process and resulting charged particles can subsequently produce significant damage to biological cells termed "direct effects". However, much of the biological damage from radiation is due to "indirect effects" from free radicals, which are the fragments of atoms that remain after being ionised. Free radicals have an unpaired or odd number of orbital electrons, and are thus chemically unstable. Such free radicals can easily break chemical bonds, and are a main cause of damage from radiation exposure.[52]

Free radicals are not unique to radiation; they are produced in response to many stressors. Damage caused from the free radicals is so abundant that efficient repair mechanisms have evolved within all biological species to counter their effects.

Radiation and the free radicals produced can damage DNA by causing several different types of lesions for which there are efficient DNA repair processes.[52] However, errors in repair can result in cell death (through apoptosis), chromosome aberrations or mutations. Mutations can be deleterious, neutral with no apparent effect (which can persist over many generations) or, rarely, may offer a selective advantage. The fate of mutations and their impacts within a population are dependent on the type of cell in which they occur. Mutations in reproductive germ cells can decrease the number of gametes, increase embryo lethality, or be inherited by the offspring, resulting in their alteration. A mutation within a somatic cell can lead to cell death, or, if DNA damaged is mis-repaired the mutation in the somatic cell can lead to cancer. The risk of non-fatal cancer for humans has been estimated at 1×10^{-5} per mSv.[52]

The deleterious effects of ionising radiation to biological systems are primarily dose dependent. The effective dose depends not only on the gross energy deposited, but also on the type of the radiation and the radiation sensitivity of the affected tissue. In SI units, the effective dose to humans is the Sievert (Sv), which is the absorbed dose (the gray; Gy) adjusted by two dimensionless weighting factors: the radiation weighting factor to account for the biological effectiveness of the absorbed radiation, and the tissue weighting factor to account for differences in the radiation sensitivities of different organs of the body. These weighting factors have been developed for human radiation biology – no such factors exist for other organisms. Thus, dose to wildlife is expressed in Gy, rather than Sv (although dose rates may be presented on a weighted or unweighted basis).

5.1 Environmental Radiological Protection

There are fundamental differences in determining the risk to humans following exposure to radiation and the risks to other organisms.[53] Human risk analyses largely focus on cancer risks to individuals. Dose–response relationships are sufficiently well known that risk factors (*i.e.* probability of lethality from cancer per unit of dose) are established. In contrast, ecological risk to wildlife is concerned generally with populations of plants and animals. For most organisms, cancer induction is not relevant and suitable endpoints include morbidity (functioning less well), reduced reproductive success, mortality and chromosomal damage. The dose–response relationships for these endpoints are not established for many wildlife groups, and therefore there are well established and quantified risk factors that equate dose to the probability of an outcome.

The endpoints considered to be most relevant in determining risks to wildlife are increased mortality, increased morbidity and decreased reproductive output. Of the three, changes in reproduction are thought to be the most sensitive to radiological exposures and relevant for the protection of wildlife populations (populations rather than individuals being likely to be the object of protection for environmental assessments).[54] Much more data are needed, however, before we can confidently predict population level impacts to wildlife as a function of radiological exposures.[55] Data are particularly scarce for the chronic, low-level exposures for which most assessments will be used. The available data on the biological effects of ionising radiation on wildlife have been compiled from the literature into an online database called the FREDERICA radiation effects database (http://www.frederica-online.org).[56]

6 Benchmarks for Wildlife Assessment

Benchmarks, or some form of usually numeric criteria, allow the outputs of environmental assessments to be placed into context and aid decisions on the need for further assessment or regulatory/remedial action. Historically, the derivation of radiological benchmarks for environmental assessment has relied upon expert judgement.

A benchmark relates to a protection goal which, in contrast to humans, is usually considered to be at the population level even though the data underpinning the benchmark relates to individuals. It can be legally binding criteria or a standard linked to a regulation where exceeding the values may result in legal or regulatory action. Alternatively, a benchmark can be a conservatively derived screening value, which aims to screen out sites where there is no cause for concern and identify those where further consideration is needed. The latter are frequently linked to tiered risk assessment schemes and serve primarily as a trigger for further investigation. Screening values in radiological assessments are often referred to as the Predicted No Effect Dose Rate (PNEDR). The methods used to derive PNEDRs are outlined in the following section; these approaches are often consistent with those used in the risk assessment of chemicals. Currently, there is no agreement on what to do if refined assessments estimate dose rates in excess of screening benchmarks.

6.1 The ICRP's Derived Consideration Reference Levels

The ICRP outlined its framework for radiological protection of the environment in ICRP Publication 108 and described its use of Reference Animals and Plants.[4] Within Publication 108, literature review and expert judgement have been used to produce Derived Consideration Reference Levels (DCRLs) for each of the RAPs (see Figure 2). The DCRLs are defined as a band of dose rate (in $mGy\,d^{-1}$) within which there is likely to be some chance of deleterious effects of ionising radiation occurring to individuals of that type of RAP

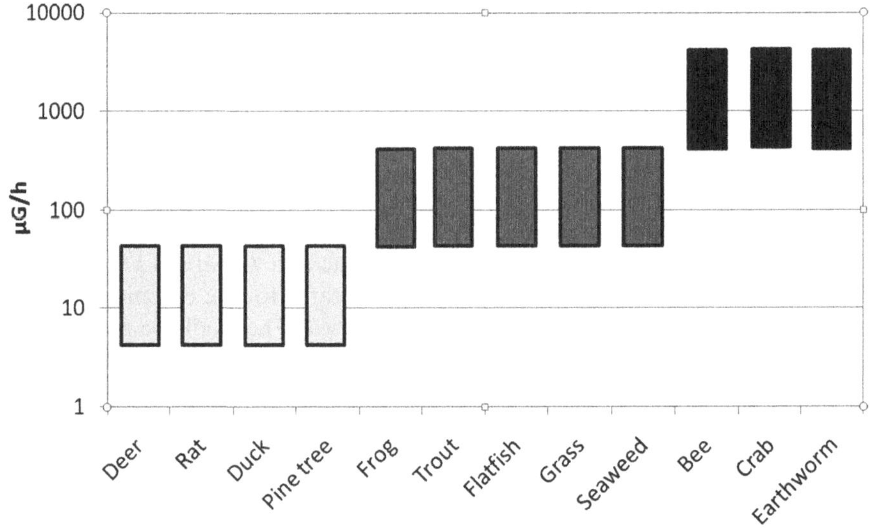

Figure 2 The ICRP Derived Consideration Reference Levels (DCRL) for Reference Animals and Plants[4] presented as $\mu Gy\,h^{-1}$.

(derived from a knowledge of expected biological effects for that type of organism) that, when considered together with other relevant information, can be used as a point of reference to optimise the level of effort expended on environmental protection, dependent upon the overall management objectives and the relevant exposure situation. The DCRLS refer to additional dose rates above that from exposure to background radionuclides (mean background dose rates from ^{238}U and ^{232}Th series radionuclides and ^{40}K are typically below 2 $\mu Gy\,h^{-1}$ for all of the ICRPs RAPs).[57,58] The ICRP is now working on guidance to show how to use the DCRLs in actual assessments.

The DCRLs varies for the RAP reflecting the differing radiosensitivities. In general, mammals, birds and pine trees are more sensitive to radiation exposure than other organisms.

6.2 Alternative Approaches used in Radiological Risk Assessments

For radiological environmental risk assessments, the benchmark may be in the form of a dose rate or back-calculated using the available assessment tools to medium environmental concentrations for each radionuclide that would give rise to the predicted no effect dose rate. These environmental concentrations [Environmental Media Concentration Limits (EMCLs) in the ERICA Tool, or Biota Concentration Guides (BCGs) in the USDOE Graded Approach] can be compared directly to measured or model-predicted environmental media concentration values and subsequently used to determine a "risk quotient". Calculated environmental media concentration benchmark values are usually applied at earlier tiers of a risk assessment for identifying (or screening out) sites where there is negligible risk of potential impact. The assumptions used in the calculation of environmental concentration benchmark values are usually conservative with respect to transfer to the organism, exposure scenario and in some instances geometry.

A risk quotient (*RQ*) provides a simple means of assessing risk by integrating the exposure and effects data to determine the likelihood of an ecological risk occurring. It is calculated from the quotient of the estimated exposure and a numeric benchmark (in the form of a dose rate or activity concentration). The benchmark dose rate is a dose rate which is assumed to be environmentally "safe". The *RQ* is defined as:

$$RQ = \frac{\textit{predicted environmental dose rate}}{\textit{benchmark dose rate assumed to be environmentally ``safe''}} \tag{9}$$

Where the resulting *RQ* is less than one, then no further effort or action would normally be required. Where the *RQ* is greater than one, then the assessment would likely need further work (such as collecting more data, refining the exposure assessment, or taking action to reduce the risk).

There are three methods commonly used to derive numeric criteria in ecotoxicology:

(i) deterministic – based on the application of assessment (or safety) factors to the most restrictive single sensitivity value observed;
(ii) probabilistic – based on Species Sensitivity Distribution (SSD) modelling; and
(iii) a weight of evidence approach – typically using data from field exposures, such as *in situ* measurements of biodiversity indices co-occurring with stressor levels.

Over the last few years, the first two approaches have been applied to radiological assessment[59–63] and are based on the guidance provided by the European Technical Guidance Document (TGD)[64] for chemical risk assessment. The benchmark produced by both approaches is designed to ensure protection of ecosystem structure and function.

The third method has not been widely used to derive benchmarks for use in radiological assessments of the environment although there are examples for specific sites (*e.g.* uranium mining).[65]

The deterministic approach, takes the lowest dose rate observed to give a significant biological effect available for any tested species and divides it by a predefined assessment/safety factor ranging from 10 to 1000 (10000 for marine ecosystems) according to the quality and quantity of the data available. The assessment/safety factor is intended to account for uncertainty and guidance on what value to apply is set out in a technical document supporting EC Directive 93/67/EEC.[64]

In contrast, the probabilistic approach uses the available (quality-assured) ecotoxicological data to determine the dose rate, giving a 10% effect resulting distribution for chronic exposure in the ecotoxicological data (the so called "effective dose rate for a 10% effect"; EDR_{10}). The EDR_{10} value is used to compensate for the influence of experimental design. For instance, the lowest no effect concentration or the highest no effect concentration may be considerably different to the true no effects concentration dependent upon experimental replication. These EDR_{10} values are then plotted together for all species for which information exists and are used to identify (usually) the fifth percentile from the species sensitivity distribution (SSD). To account for any residual uncertainty, an assessment factor of between 1 and 5 is applied to the fifth percentile value based on the available quality and quantity of the data in the SSD to produce the predicted no effect dose rate. This approach, as applied to radiological assessment, is described in full by Garnier-Laplace *et al.*[60–62]

The SSD approach has been used to derive a screening dose rate of $10\,\mu Gy\,h^{-1}$ using different data selection criteria;[59,60] this value is used as the default screening dose rate in the ERICA Tool.[13] The screening dose rate is to be applied to incremental (*i.e.* above background) exposure. Currently, it has only been possible to derive a generic screening value applied to all ecosystems using this approach due to the lack of appropriate quantitative data across a sufficient number of different wildlife categories. Thus, it is not possible to derive screening values by the SSD approach sub-divided into different wildlife groups due to statistical constraints.

A screening dose rate is for application in regulatory assessments of planned releases and is not a useful benchmark for use in accidental situations such as that ongoing at Fukushima. Consideration of effects on wildlife after the Chernobyl accident with reference to Fukushima is presented by Beresford and Copplestone.[70]

Acknowledgements

The authors have used some materials prepared for a NERC funded course on radiation protection in the environment (http://www.ceh.ac.uk/protect). These notes were prepared by Jordi Vives i Batlle (SCK·CEN), Tom Hinton (IRSN) and David Copplestone (Stirling University). We are grateful to the authors for allowing the use of these materials in this chapter. We would also like to thank Claire Wells and Cath Barnett (CEH) for their assistance during the preparation of this manuscript.

References

1. D. Copplestone, J. E. Brown and N. A. Beresford, *J. Radiol. Prot.*, 2010, **30**, 283–297.
2. ICRP, 1990 *Recommendations of the International Commission on Radiological Protection*, ICRP Publication 60, Ann. ICRP 21(1–3), Elsevier, 1990.

3. ICRP, 2007 *Recommendations of the International Commission on Radiological Protection*, ICRP Publication 103, Ann. ICRP 37(2–4), Elsevier, 2007.
4. ICRP, *Environmental Protection: The Concept and Use of Reference Animals and Plants*, ICRP Publication 108, Ann. ICRP 38(4–6), Elsevier, 2009.
5. IAEA, *International Basic Safety Standards for Protection against Ionizing Radiation and for the Safety of Radiation Sources*, Safety Series No. 115, IAEA, Vienna, Austria, 1996.
6. IAEA, *Fundamental Safety Principles*, IAEA Safety Standards Series No SF-1 IAEA, Vienna, Austria, 2006.
7. Environment Canada, *Environmental Assessments of Priority Substances under the Canadian Environmental Protection Act*, Guidance Manual Version 1.0 – March 1997, EPS 2/CC/3E, Environment Canada, Ottawa, ON, Canada, 1997.
8. Strålsäkerhetsmyndigheten, *Strålsäkerhetsmyndighetens föreskrifter och allmänna råd om skydd av människors hälsa och miljön vid slutligt omhändertagande av använt kärnbränsle och kärnavfall*, SSMFS 2008:37, ISSN 2000-0987, The Swedish Radiation Protection Institute's regulations on the protection of human health and the environment in connection with the final management of spent nuclear fuel and nuclear waste, 2009.
9. USDOE, *A Graded Approach for Evaluating Radiation Doses to Aquatic and Terrestrial Biota*, Technical standard DOE-STD-1153-2002, Modules 1-3, US Department of Energy, Washington DC, USA, 2002.
10. USDOE, *Environmental Protection Program*, DOE Order 450.1A, US Department of Energy, Washington DC, 2008.
11. D. Copplestone, M. D. Wood, S. Bielby, S. R. Jones, J. Vives and N. A. Beresford, *Habitat Regulations for Stage 3 Assessments: Radioactive Substance Authorisations*, Environment Agency R&D Technical Report P3-101/SP1a, Bristol, UK, 2003, p. 100.
12. D. Copplestone, M. D. Wood, P. C. Merrill, R. Allott, S. R. Jones, J. Vives, N. A. Beresford and I. Zinger, *Radioprotection*, 2005, **40**, S893–S898.
13. J. E. Brown, B. Alfonso, R. Avila, N. A. Beresford, D. Copplestone, G. Pröhl and A. Ulanovsky, *J. Environ. Radioact.*, 2008, **99**, 1371–1383.
14. D. Copplestone, S. Bielby, S. R. Jones, D. Patton, P. Daniel and I. Gize, *Impact Assessment of Ionising Radiation on Wildlife*, R&D Publication 128, Environment Agency, Bristol, UK, 2001.
15. C.-M. Larsson, *J. Radiol. Prot.*, 2004, **24**, A1–A13.
16. N. C. Garisto, F. Cooper and S. L. Fernandes, *No-effect Concentrations for Screening Assessment of Radiological Impacts on Non-human Biota*, NWMO TR-2008-02, Nuclear Waste Management Organisation, Toronto, ON, Canada, 2008.
17. C.-M. Larsson, *J. Environ. Radioact.*, 2008, **99**, 1364–1370.
18. USDOE, *RESRAD-BIOTA: A Tool for Implementing a Graded Approach to Biota Dose Evaluation*, user's guide, version 1, ISCORS technical report 2004-02 *DOE*/EH-0676, 2004.
19. G. Pröhl, *J. Environ. Radioact.*, 2008, **100**, 675–682.
20. E. Leclerc and Y. H. Choi, Weathering, in *International Atomic Energy Agency, Quantification of Radionuclide Transfer in Terrestrial and*

Freshwater Environments for Radiological Assessments, IAEA-TECDOC-1616, IAEA, Vienna, Austria, 2009.

21. C. J. Gil-Garcia, A. Rigol and M. Vidal, *J. Environ. Radioact.*, 2009, **100**, 690–696.
22. M. Frissel, H. Noordijk and K. E. Van Bergeijk, in *Transfer of Radionuclides in Natural and Semi-natural Environments*, ed. G. Desmet, S. Nassimbeni and M. Belli, Elsevier, London, UK, 1990, pp. 40–47.
23. S. M. Wright, J. T. Smith, N. A. Beresford and W. A. Scott, *Radiat. Environ. Biophys.*, 2003, **42**, 41–47.
24. A. Sanchez, S. M. Wright, E. Smolders, C. Naylor, P. A. Stevens, V. H. Kennedy, B. A. Dodd, D. L. Singleton and C. L. Barnett, *Environ. Sci. Technol.*, 1999, **33**, 2752–2757.
25. J. Absalom, S. D. Young, N. M. J. Crout, A. L. Sanchez, S. M. Wright, E. Smolders, A. F. Nisbet and A. G. Gillett, *J. Environ. Radioact.*, 2001, **52**, 31–43.
26. N. Sanzharova, S. Fesenko and E. Reed, *Processes governing radionuclide transfer to plants, in International Atomic Energy Agency Quantification of Radionuclide Transfer in Terrestrial and Freshwater Environments for Radiological Assessments*, IAEA-TECDOC-1616, IAEA, Vienna, Austria, 2009.
27. E. IAEA, *Quantification of Radionuclide Transfer in Terrestrial and Freshwater Environments for Radiological Assessments,* IAEA-TECDOC-1616, IAEA, Vienna, Austria, 2009.
28. IAEA, *Handbook of Parameter Values for the Prediction of Radionuclide Transfer in Terrestrial and Freshwater Environments*, TRS 472, IAEA, Vienna, Austria, 2010.
29. N. A. Beresford, R. W. Mayes, A. I. Cooke, C. L. Barnett, B. J. Howard, C. S. Lamb and G. P. L. Naylor, *Environ. Sci. Technol.*, 2000, **34**, 4455–4462.
30. ICRP, *Human Alimentary Tract Model for Radiological Protection*, ICRP Publication 100, Ann. ICRP 36(1–2), 2006.
31. B. J. Howard, N. A. Beresford, C. L. Barnett and S. Fesenko, *J. Environ. Radioact.*, 2009, **100**, 1069–1078.
32. N. A. Beresford, R. W. Mayes, C. L. Barnett and B. J. Howard, *J. Environ. Radioact.*, 2007, **98**, 24–35.
33. A. M. Galer, N. M. J. Crout, N. A. Beresford, B. J. Howard, R. W. Mayes, C. L. Barnett, H. F. Eayres and C. S. Lamb, *J. Environ. Radioact.*, 1993, **20**, 35–48.
34. N. A. Beresford, N. M. J. Crout, R. W. Mayes, B. J. Howard and C. S. Lamb, *J. Environ Radioact.*, 1998, **38**, 317–338.
35. N. M. J. Crout, N. A. Beresford, B. J. Howard, R. W. M. Mayes and H. S Hansen, *J. Dairy Sci.*, 1998, **81**, 92–99.
36. N. M. J. Crout, R. W. Mayes, N. A. Beresford, C. S. Lamb and B. J. Howard, *Radiat. Environ. Biophys.*, 1998, **36**, 243–250.
37. N. A. Beresford, C. L. Barnett, J. Brown, J.-J. Cheng, D. Copplestone, V. Filistovic, A. Hosseini, B. J. Howard, S. R. Jones, S. Kamboj, A. Kryshev, T. Nedveckaite, G. Olyslaegers, R. Saxén, T. Sazykina, J. Vives I Batlle, S. Vives-Lynch, T. Yankovich and C. Yu, *Radiat. Environ. Biophy.*, 2008, **47**, 491–514.

38. N. A. Beresford, C. L. Barnett, K. Beaugelin-Seiller, J. E. Brown, J.-J. Cheng, D. Copplestone, S. Gaschak, J. L. Hingston, J. Horyna, A. Hosseini, B. J. Howard, S. Kamboj, A. Kryshev, T. Nedveckaite, G. Olyslaegers, T. Sazykina, J. T. Smith, D. Telleria, J. Vives i Batlle, T. L. Yankovich, R. Heling, M. D. Wood and C. Yu, *Radioprotection*, 2009, **44**, 565–570.
39. N. A. Beresford, C. L. Barnett, J. E. Brown, J.-J. Cheng, D. Copplestone, S. Gaschak, A. Hosseini, B. J. Howard, S. Kamboj, T. Nedveckaite, G. Olyslaegers, J. T. Smith, J. Vives I Batlle, S. Vives-Lynch and C. Yu, *J. Radiol. Prot.*, 2010, **30**, 341–373.
40. T. L. Yankovich, J. Vives i Batlle, S. Vives-Lynch, N. A. Beresford, C. L. Barnett, K. Beaugelin-Seiller, J. E. Brown, J.-J. Cheng, D. Copplestone, R. Heling, A. Hosseini, B. J. Howard, S. Kamboj, A. I. Kryshev, T. Nedveckaite, J. T. Smith and M. D. Wood, *J. Radiol. Prot.*, 2010, **30**, 299–340.
41. IAEA, *Handbook of Parameter Values for the Prediction of Radionuclide Transfer to Wildlife*, IAEA, Vienna, in preparation.
42. T. L. Yankovich, N. A. Beresford, M. Wood, T. Aono, P. Andersson, C. L. Barnett, P. Bennett, J. Brown, S. Fesenko, A. Hosseini, B. J. Howard, M. Johansen, M. Phaneuf, K. Tagami, H. Takata, J. Twining and S. Uchida, *Radiat. Environ. Biophys.*, 2010, **49**, 549–565.
43. N. A. Beresford, C. L. Barnett, B. J. Howard, W. A. Scott, J. E. Brown and D. Copplestone, *J. Environ. Radioact.*, 2008, **99**, 1393–1407.
44. K. A. Higley, *Radiat. Environ. Biophys.*, 2010, **49**, 657–672.
45. J. M. Gómez-Ros, G. Pröhl, A. Ulanovsky and M. Lis, *J. Environ. Radioact.*, 2008, **99**, 1449–1455.
46. J. Vives i Batlle, M. Balonov, K. Beaugelin-Seiller, N. A. Beresford, J. Brown, J.-J. Cheng, D. Copplestone, M. Doi, V. Filistovic, V. Golikov, J. Horyna, A. Hosseini, B. J. Howard, S. R. Jones, S. Kamboj, A. Kryshev, T. Nedveckaite, G. Olyslaegers, G. Pröhl, T. Sazykina, A. Ulanovsky, S. Vives Lynch, T. Yankovich and C. Yu, *Radiat. Environ. Biophys.*, 2007, **46**, 349–373.
47. N. A. Beresford, A. Hosseini, J. E. Brown, C. Cailes, K. Beaugelin-Seiller, C. L. Barnett and D. Copplestone, *J. Radiol. Prot.*, 2010, **30**, 265–284.
48. D. B. Chambers, R. V. Osborne and A. L Garva, *J. Environ. Radioact.*, 2006, **87**, 1–14.
49. A. Ulanovsky and G. Pröhl G, *Radiat. Environ. Biophys.*, 2006, **45**, 203–214.
50. J. Vives i Batlle, K. Beaugelin-Seiller, N. A. Beresford, D. Copplestone, J. Horyna, A. Hosseini, M. Johansen, S. Kamboj, D.-K. Keum, N. Kurosawa, L. Newsome, G. Olyslaegers, H. Vandenhove, S. Ryufuku, S. Vives Lynch, M. D. Wood and C. Yu, *Radiat. Environ. Biophys.*, 2010, available online (DOI: 10.1007/s00411-010-0346-5).r.
51. F. W. Whicker and V. Schultz, *Radioecology: Nuclear Energy and the Environment,* CRC Press, Boca Raton, FL, USA, 1982.
52. IAEA, *Radiation Biology: A Handbook for Teachers and Students*, IAEA-TCS-42. ISSN 1018-5518, IAEA, Vienna, Austria, 2010.
53. T. G. Hinton, Risks from exposure to radiation, in *Fundamentals of Ecotoxicology*, ed. M. C. Newman, Ann Arbor Press, Chelsea, ISBN-1575040131, 1998, ch. 14.

54. IAEA, *Effects of Ionizing Radiation on Plants and Animals at Levels Implied by IAEA,* IAEA, Vienna, Austria, 1992.
55. T. G. Hinton, J. S. Beford, J. C. Congdon and F. W. Whicker, *Radiat. Res.*, 2004, **162**, 332–338.
56. D. Copplestone, J. L. Hingston and A. Real, *J. Environ. Radioact.*, 2008, **99**, 1456–1463.
57. N. A. Beresford, C. L. Barnett, D. G. Jones, M. D. Wood, J. D. Appleton, N. Breward and D. Copplestone, *J. Environ. Radioact.*, 2008, **99**, 1430–1439.
58. A. Hosseini, N. A. Beresford, J. E. Brown, D. G. Jones, M. Phaneuf, H. Thørring and T. Yankovich, *J. Radiol. Prot.*, 2010, **30**, 235–264.
59. P. Andersson, J. Garnier-Laplace, N. A. Beresford, D. Copplestone, B. J. Howard, P. Howe, D. Oughton and P. Whitehouse, *J. Environ. Radioact.*, 2009, **100**, 1100–1108.
60. J. Garnier-Laplace, C. Della-Vedova, R. Gilbin, D. Copplestone, J. L. Hingston and P. Ciffroy, *Environ. Sci. Technol.*, 2006, **40**, 6498–6505.
61. J. Garnier-Laplace, D. Copplestone, R. Gilbin, F. Alonzo, P. Ciffroy, M. Gilek, A. Aguero, M. Bjork, D. Oughton, A. Jaworsk, C. M. Larsson and J. Hingston, *J. Environ. Radioact.*, **99**, 1474–1483.
62. J. Garnier-Laplace, C. Della-Vedova, P. Andersson, D. Copplestone, C. Cailes, N. A. Beresford, B. J. Howard, P. Howe and P. Whitehouse, *J. Radiol. Prot.*, 2010, **30**, 215–233.
63. J. R. Twining, J. M. Ferris, I. Zinger and D. Copplestone, in *Protection of the Environment from the Effects of Ionizing Radiation*, Contributed Papers, STI/PUB/1229 Companion CD, IAEA CN-109/114. IAEA, Vienna, 2005, 3–11.
64. EC, *Technical Guidance Document in Support of Commission Directive 93/67/EEC on Risk Assessment for New Notified Substances*, Commission Regulation (EC) No. 1488/94 on Risk assessment for existing substances and Directive 98/8/EC of the European Parliament and of the Council concerning the placing of biocidal products on the market, Office for Official Publication of the European Commission, Luxembourg, 2003.
65. P. A. Thompson, J. Kurias and S. Mihok, *Environ. Monit. Assess.*, 2005, **100**, 71–85.
66. Ministry of Health, Labour and Welfare, 2011, *Information about the great East Japan Earthquake, Food*, http://www.mhlw.go.jp/english/topics/2011eq/index.html.
67. N. A. Beresford and B. J. Howard, An overview of the transfer of radionuclides to farm animals and potential countermeasures of relevance to Fukushima releases, Integrated Environmental Assessment and Management, in press.
68. B. J. Howard, N. A. Beresford, C. L. Barnett and S. Fesenko, *J. Environ. Radioact.*, 2009, **100**, 263–273.
69. B. J. Howard, N. A. Beresford, C. L. Barnett and S. Fesenko, *J. Environ. Radioact.*, 2009, **100**, 767–773.
70. N. A. Beresford and D. Copplestone, *Intergrated Environmental Assessment and Management*, in press.

Radiological Protection of Workers and the General Public

JAN PENTREATH

ABSTRACT

Radiological protection has a long pedigree; its origins go back almost a century. Since then, not only has a large body of information on the effects of radiation been accrued, but this information has been used in a most successful way to manage the exposures of people in all forms of exposure situations. This success has largely been due to a unique organization, the International Commission on Radiological Protection (ICRP), that has continually evaluated, interpreted and worked out how best to apply the knowledge that has arisen from the scientific disciplines of radiation physics, dosimetry and radiobiology, together with the complicated interpretation of numerous epidemiological studies. Furthermore, the ICRP has attempted to interface the continually improving science with the ever changing cultural and sociological context within which radiological protection needs to be applied. This is an on-going task. The current situation is one in which radiological protection guidance is set out within a framework of three exposure situations (planned, emergency and existing), involving three categories of human exposure (medical, occupational and public). Due to the scientific interpretation of the data, this matrix of exposure situations, and categorization of those likely to be exposed, is handled within a set of principles of justification of exposure, the optimization of the level of protection and the application of dose limits. All of the elements of this framework are briefly set out and discussed in this chapter, together with a brief overview of the current rates of exposure, due to different exposure situations, for people within the UK.

Issues in Environmental Science and Technology, 32
Nuclear Power and the Environment
Edited by R.E. Hester and R.M. Harrison

Published by the Royal Society of Chemistry, www.rsc.org

1 Introduction

There is probably more known about the effects of radiation on human beings than there is about the effects of any other hazard relating to the generation of energy on a large scale. In addition, this knowledge is better organized, managed and converted into useful practical procedures at an internationally agreed level, than anything comparable in relation to other hazards. The reasons for this lie in the fact that an enormous data base has been accrued over almost a century, and because of the establishment of what is a unique and, by now, a somewhat ancient body – the International Commission on Radiological Protection (ICRP).

Its origins lie in the fact that, following their discovery at the end of the 19th century, the medical benefits of X-rays, and of the gamma rays from radium, were very quickly recognized. The dangers of radiation soon became apparent, however, and national committees set up to address the problems started to appear in 1913. There was also a need for some form of international co-operation, and following a decision by the Second International Congress of Radiology, the ICRP was established in 1928 under the name of the International X-Ray and Radium Protection Committee. It was restructured in 1950 and given its present name.

The ICRP is an advisory body. It regularly issues detailed advice and information regarding protection against the hazards of ionizing radiation and, at suitable intervals, revises its overall set of "Recommendations". The first report in the current publication series contained the Recommendations that had been adopted in 1958[1] and, since then, revisions have been set out in Publication 26,[2] Publication 60[3] and, most recently, in Publication 103.[4]

The advice of the ICRP is aimed principally at regulatory authorities, organizations, and individuals that have responsibility for radiological protection, and virtually all international standards and national regulations addressing radiological protection are based on its recommendations. There is a close connection between the Recommendations and the International Basic Safety Standards for Protection against Ionizing Radiation and the Safety of Radiation Sources (usually simply called "the BSS"), which are co-sponsored by the relevant international organisations within the UN family and issued by the International Atomic Energy Agency (IAEA). The governing body of the IAEA requires that the BSS take the ICRP's Recommendations into account. These Recommendations are then, in turn, cascaded down to such bodies as the OECD's Nuclear Energy Agency (NEA), and to regional bodies (such as EURATOM), and to national bodies, such as what was the UK's National Radiological Protection Board, the functions of which are now part of the Health Protection Agency (HPA).

The ICRP operates *via* a set of five committees, each of which also makes extensive use of specialized task groups. It also works closely with the United Nations Scientific Committee on the Effects of Atomic Radiation (UNSCEAR) that was established in 1955 by the UN General Assembly with a mandate to assess and report on levels of, and exposure to, ionizing radiation. The ICRP thus draws upon a vast amount of world-wide experience and data bases on all forms of radiation exposure relating to medical practices, industrial workers and the general public, and, more recently, on the natural environment as well.

The first dose unit, the roentgen (R), was established for X-rays in 1928 by yet another committee, the International X-ray Unit Committee, which was later to become the International Commission on Radiation Units (ICRU). The first official use of the term "dose", together with an amended definition of the unit R, appeared in the 1937 recommendations of the ICRU.[5] The ICRU raised the concept of an absorbed dose, and officially defined the name and its unit, the "rad", in 1953 to extend the concept of dose to certain materials other than air.[6] The first dose quantity incorporating the concept of different responses of tissues to different types of radiation (known as "relative biological effectiveness", or RBE), was the "RBE dose in rems" (rem stood for roentgen equivalent in man, or mammal). This was an RBE-weighted sum of absorbed dose in rads, proscribed in the 1956 recommendations of the ICRU. As a result of joint efforts between the ICRU and the ICRP, it was later replaced by the "dose equivalent" which was defined by the product of absorbed dose, the quality factor of the radiation, the dose distribution factor and other necessary modifying factors.[7] The "rem" was retained as the unit of dose equivalent. The ICRU also defined another dose quantity, "kerma", and changed the name of "exposure dose" to the simple one of "exposure" in its 1962 recommendations.

Since then much has happened, including the introduction of SI units. The fundamental dosimetric quantity in radiological protection is now the *absorbed dose* (D). This is the energy absorbed per unit mass, and its unit is the joule per kilogram, which is given the special name gray (Gy). Absorbed dose is defined in terms that allow it to be specified at a point, but it is used by the ICRP, except where otherwise stated, to mean the average dose over a tissue or organ. Multiplying the absorbed dose by appropriate weighting factors, depending on the type of radiation, creates the *equivalent dose* (H_T) in the relevant organ or tissue. The equivalent dose is preferred in radiation protection because it is more closely related to the *risk* of harm in the exposed organ or tissue. By weighting the equivalent dose in each organ in proportion to its radiation sensitivity (in other words, to the probability and severity of the harm done by radiation), and then adding the weighted contributions from each organ to a total body dose, a third dose, the *effective dose* (E), is obtained. In radiation protection it is usually the effective dose that is determined for comparison with dose limits or for assessments of risks. Both the equivalent dose and the effective dose are measured in a unit called the sievert (Sv). For some applications, a collective dose may be calculated, being the product of the number of exposed individuals and their average dose. The *collective effective dose* (S) may also sometimes be used as a measure of the expected collective harm, sometimes referred to as the "radiation health detriment".

Thus a system has been developed where the science base relating exposure to dose, and dose to effects, is examined, re-examined and interpreted by way of a set of conceptual and numeric "model" (see Figure 1). This modelling process

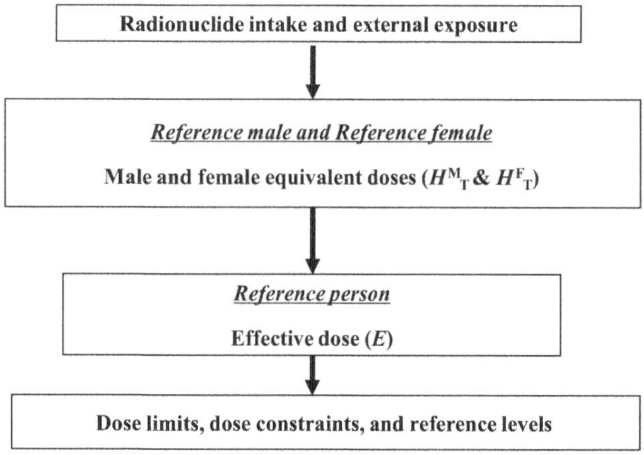

Figure 1 Steps in the derivation of numerical advice for the protection of people.[9]

started with the creation of a "Reference Man", who has now evolved into a Reference Individual (male and female), and a Reference Person. The former is an idealised male or female entity with reference anatomical and physiological characteristics, as defined by ICRP.[8] Thanks to the rapid development of the relevant technology, phantoms based on medical tomographic images, consisting of three-dimensional volume pixels (voxels), are now used to compute the mean absorbed dose in an organ or tissue, and these doses are multiplied by radiation weighting factors to provide the equivalent doses in the Reference Male and Reference Female. These steps are shown in Figure 1.

For the purposes of radiological protection, however, it is currently thought useful to apply a single value of *effective* dose irrespective of sex. This is achieved by deriving sex-averaged organ or tissue equivalent doses for an idealised Reference Person and then multiplying them by the corresponding tissue weighting factors.

2 The Health Effects of Radiation

This sophisticated approach would not be possible were it not for the fact that there is sufficient information on the effects of radiation on humans to be able to differentiate between levels of dose received by different tissues and organs from both external and internal sources, and the different types of effect that result. These relationships are complex, but in order to *manage* the exposure of people to radiation, it is useful to consider the principal effects of radiation as being those that result in either:

(i) deterministic effects, which are due in large part to the killing or malfunction of cells following high doses of exposure to radiation; and those that cause

(ii) stochastic effects, such as cancer and heritable effects, which involve either cancer development in exposed individuals due to the mutation of somatic cells, or heritable disease in their offspring owing to mutation of reproductive (germ) cells.

These two categories do not of course cover every adverse health effect, and thus consideration is also given to effects on the embryo and foetus, and to diseases other than cancer.

The induction of deterministic effects (or what are also called "tissue reactions") is generally characterized by a threshold dose of radiation, because radiation damage to a critical population of cells, in a given tissue, needs to be sustained before injury is expressed in a clinically relevant form. Above the threshold dose, the severity of the injury, including impairment of the capacity for tissue recovery, increases with increasing dose. In cases where the threshold dose has been exceeded, early tissue reactions (days to weeks) may be of an inflammatory type, resulting from the release of cellular factors, or there may be reactions resulting from cell loss. Late tissue reactions (months to years) can be of a generic type if they arise as a direct result of damage to that tissue. Essentially, in the absorbed dose below about 100 mGy, tissues are judged generally not to express clinically relevant functional impairment, and this judgement applies both to single acute doses, and to situations where these low doses are experienced in a protracted form as repeated annual exposures.

With regard to stochastic effects, epidemiological and experimental studies provide evidence of radiation risks of cancer, albeit with uncertainties, at doses of about 100 mSv or less. As to be expected, a very large amount of effort has been expended on trying to understand, and quantify, this cancer risk. As far as the mechanisms are concerned, the accumulation of cellular and animal data over decades of study lead to the view that DNA damage response processes within single cells are of critical importance to the development of cancer after radiation exposure. Of particular importance are effects such as the induction of complex forms of DNA double strand breaks, the problems experienced by cells in correctly repairing these complex forms of DNA damage, and the consequent appearance of gene or chromosomal mutations.

Due to its stochastic nature, great reliance is necessarily placed on epidemiological information relating to the incidence of cancer. These data bases are continually growing, and a lot of information on the risk of organ-specific cancer following exposure to radiation has come from the continuing follow-up of survivors of the 1945 atomic bomb explosions in Japan – the so-called Life Span Study (LSS). These data relate to both cancer mortality and cancer incidence, the latter providing more reliable estimates of risk. In addition, epidemiological data from the LSS provide further information on the temporal and age-dependent pattern of radiation cancer risk, particularly the assessment of risk amongst those exposed at an early age.

The LSS is not, however, the sole source of information, and data from medical, occupational and environmental studies are also considered and evaluated. For cancers occurring in some parts of the human body, there is reasonable compatibility between the data from the LSS and those from other sources. There are also differences in radiation risk estimates among the various data sets, however, and most studies on environmental radiation exposures currently lack sufficient data (on dosimetry and tumour ascertainment) to contribute directly to risk estimation.

Part of the problem of relating the cancer risk to radiation exposure is the large range of doses and dose rates over which observations have been made. With low doses the risk of developing cancer is much less, and thus the data bases need to be much larger in order to obtain statistically significant information. A dose and dose-rate effectiveness factor (DDREF) has therefore been used by UNSCEAR to project cancer risks determined at high doses, and high dose rates, to the risks that would apply at low doses, and low dose rates. From a combination of epidemiological, animal and cellular data, cancer risks at low doses and low dose rates are judged, in general, to be reduced by the value of the factor ascribed to the DDREF. And although, in reality, different dose and dose rate effects may well apply to different organs and tissues, the ICRP judges that, for the general purposes of radiological protection, a DDREF of 2 should be applied, and this is used to derive its nominal risk coefficients for all cancers.

One approach that has been used to manage cancer risk at low doses and low dose rates is to assume that, at doses below about 100 mSv, a given increment in dose will produce a directly proportional increment in the probability of incurring cancer (or heritable effects) attributable to radiation. This dose–response model is generally known as the "linear-non-threshold" or LNT model. It is a view that accords with that given by UNSCEAR,[10] but other estimates have been provided by various national organisations. Some of these are in line with the UNSCEAR view[11,12] whilst others are not – such as the report from the French Academies,[13] which argues in support of a practical threshold for radiation cancer risk.

As far as the ICRP is concerned, although it recognises that there are exceptions, it judges that the weight of evidence supports the view that, *for the purposes of radiological protection*, it is scientifically plausible to assume that the incidence of cancer or heritable effects will rise in direct proportion to an increase in the equivalent dose in the relevant organs and tissues at doses below about 100 mSv. This is, of course, a practical judgement. In arriving at it, the potential challenges associated with information on cellular adaptive responses, and on the relative abundance of spontaneously arising and low-dose-induced DNA damage, have also been considered. Indeed, there are other factors to consider, such as the existence of the post-irradiation cellular phenomena of induced genomic instability and bystander signalling. All of these biological factors, together with possible tumour-promoting effects of protracted irradiation and immunological phenomena, may influence radiation cancer risk.[14] Because the estimation of *nominal* cancer risk coefficients is based upon direct human epidemiological data, however, any contribution from them would, in any case, be included in that estimate.

One further point in relation to the LNT model needs to be noted, and that is whilst it remains a scientifically plausible element in a practical system of radiological protection, any biological or epidemiological information that would, unambiguously, verify the hypothesis that underpins the model is unlikely to be forthcoming.[10,11] Due to this uncertainty, the ICRP advises that it is *not* appropriate, for the purposes of public health planning, to calculate the hypothetical number of cases of cancer or heritable disease that might be associated with very small radiation doses, received by large numbers of people, over very long periods of time.

In addition, contrary to the commonly held view that often appears in the popular press, there is currently no direct evidence that the exposure of human parents to radiation leads to an excess of heritable disease in their offspring. There is compelling evidence, however, that radiation does cause heritable effects in experimental animals. The ICRP therefore prudently includes the risk of heritable effects in its system of radiological protection. This risk is based on the concept of the doubling dose (DD) for disease-associated mutations, for which the present estimate, up to the second generation, is about 0.2% per Gy.

With regard to the embryo and foetus, it is recognized that there is a susceptibility to the effects of irradiation in the pre-implantation period of embryonic developments, and there are gestational age-dependent patterns of *in utero* radiosensitivity, with maximum sensitivity being expressed during the period when organs are being formed. At doses under 100 mGy, lethal effects are very infrequent and, on the basis of animal data, it is judged that there is a true dose threshold of around 100 mGy for the induction of malformations. For practical purposes, therefore, it is assumed that risks of malformation after *in utero* exposure to doses well below 100 mGy would not be expected. With regard to cancer risk following *in utero* irradiation, the largest case-control studies of *in utero* (medical) irradiation provide evidence that there is an increased risk of childhood cancer of all types. There are particular uncertainties regarding the risk of radiation-induced solid cancers following *in utero* exposure, but it is prudent to assume that life-time cancer risk following *in utero* exposure will be similar to that following irradiation in early childhood which is, at most, about three times that of the population as a whole.

3 The Scientific Framework for the Protection of Humans

The scientific framework that has evolved for the protection of humans from ionizing radiation is based on a number of related features, including the use of reference anatomical and physiological models for the assessment of radiation doses from external and internal sources; studies of radiation effects at the molecular and cellular level; a large range of experimental animal studies; plus epidemiological studies of exposed populations over many decades. Models and data have been used to derive tabulated, standardised data on the committed "dose per unit intake" of different radionuclides for internal exposures, and "dose per unit air kerma or fluence" for external exposures of workers, patients and the public. Epidemiological and experimental studies have been used in the estimation of risks associated with external and internal radiation exposure. For biological effects, the data from human experience have been further supported by experimental biology. For cancer, and for heritable effects, the starting points are the results of epidemiological studies and of studies on animal and human genetics. These results are, in turn, supplemented by information from experimental studies on the mechanisms of carcinogenesis and heredity, in order to provide risk estimates at the low doses of interest in radiological protection.

In interpreting these data, certain balances have to be struck. With regard to radiation weighting factors, those for photons, electrons and muons are assigned a radiation weighting factor of 1. This is a simplification, particularly for photons, but is considered sufficient for their use in equivalent and effective dose terms because these are used for dose limitation, and assessment and control, in the low dose range. With regard to protons, external radiation sources are of most concern, and a radiation factor of 2 is used. A factor of 2 is also used for pions. These are particles of importance for exposures in aircraft, and for those involved with high-energy particle accelerators. Alpha particles are particularly important with regard to internal exposures, and a weighting value of 20 is used. A value of 20 is also used for fission fragments, which are also of importance with regard to internal exposures, and the same value is used for heavy ions, which are encountered in high altitude aviation and space exploration. Finally, neutrons are treated somewhat differently, and the radiation weighting factor for them differs in relation to energy over a range of about 2.5 to slightly over 20.

Similarly, balances have to be struck in view of the uncertainties surrounding the values of tissue weighting factors and the estimate of detriment. Thus it is currently considered appropriate, again for radiological protection purposes, to use age and sex averaged tissue weighting factors and numerical risk estimates. For stochastic effects, after exposure to radiation at low dose rates, nominal probability coefficients for detriment-adjusted cancer risk of 5.5×10^{-2} Sv^{-1} for the whole population and 4.1×10^{-2} Sv^{-1} for adult workers have been derived. For heritable effects, the detriment-adjusted nominal risk in the whole population is estimated at 0.2×10^{-2} Sv^{-1} and in adult workers at 0.1×10^{-2} Sv^{-1}.

These risk estimates are called "nominal" by the ICRP because they relate to the exposure of a nominal population of males and females, with a typical age distribution, and are computed by averaging over age groups and both sexes. The dosimetric quantity recommended for radiological protection, the *effective dose*, is also computed by age and sex averaging. There are many uncertainties inherent in the definition of nominal factors to assess effective dose, but the estimates of fatality and detriment coefficients are considered adequate for radiological protection purposes. Nevertheless, as with all estimates derived from epidemiology, the nominal risk coefficients do not of course apply to specific individuals. For the estimation of the likely consequences of an exposure of a known individual, or of a known population, it is necessary to use specific data relating to that exposed individual or population.

In those situations in which the dose thresholds (100 mSv in a year) for deterministic effects in relevant organs could be exceeded, protective actions should be taken. At radiation doses below around 100 mSv in a year, the increase in the incidence of stochastic effects is assumed to occur with a small probability, and in proportion to the increase in radiation dose over the background dose.

In terms of managing exposure to radiation, it is also necessary to consider that, on the one hand, individuals may be simultaneously exposed to several sources, and thus an "individual-related" assessment of the total exposure has to be attempted; whereas, on the other hand, it is also necessary to consider all of the individuals exposed by a single radiation source or group of sources – a "source-related" assessment. Of the two, the primary importance, of course, is the source-related assessment, because action can be taken for a source to assure the protection of all individuals from that source.

In terms of presenting the scientific framework of radiological protection to a wider audience, the probabilistic nature of stochastic effects and the properties of the LNT model make it impossible to derive a clear distinction between "safe" and "dangerous". This inevitably creates some difficulties in explaining the control of radiation risks. The major policy implication of the LNT model is that some finite risk, however small, must be assumed, and a level of protection therefore has to be established that is based upon what is deemed acceptable by society at any one time. This is the problem that the ICRP's system of protection attempts to address.

4 The ICRP's System of Protection

The primary aim of the ICRP's Recommendations is to contribute to an appropriate level of protection for people and the environment against the detrimental effects of radiation exposure, without unduly limiting the desirable human actions that may be associated with such exposure. In protecting individuals, it is the control (in the sense of restriction) of radiation doses that is important, no matter what the source. In view of what is known about the effects of radiation, the human health objectives are relatively straightforward: to manage and control exposures to ionising radiation so that deterministic effects are prevented and the risks of stochastic effects are reduced to the extent reasonably achievable. Before examining how these objectives are achieved, however, it is first useful to consider the situations that would result in radiation exposure in the first place.

The ICRP currently recognises three types of exposure situations:

(i) *Planned* exposure situations, which are situations involving the planned introduction and operation of sources, and include situations that were previously categorised as "practices". These include situations that are anticipated to occur (in other words, "normal" exposures) as well as exposures that are not anticipated to occur but may occur ("potential" exposures), such as accidents. In the latter case, although the situation was not planned to occur, the situation itself can be planned for, although not necessarily in great detail. These days such potential exposures can include a variety of possibilities, from accidents that may occur unintentionally to those where there is a clear intention to perform a malevolent act. Specific guidance has been given in relation to radiological attacks.[15]

(ii) *Emergency* exposure situations, which are unexpected situations such as those that may occur during the operation of a planned situation, requiring urgent attention. They are, inevitably, unpredictable in detail, and often require particular attention being paid to deterministic health effects.

(iii) *Existing* exposure situations, which are exposure situations that already exist when a decision on control has to be taken, such as those caused by natural background radiation.

Individuals may be exposed to radiation from more than one source. Provided that doses are below the threshold for deterministic effects (harmful tissue reactions), the presumed proportional relationship between the additional dose attributable to each situation and the corresponding increase in the probability of stochastic effects makes it possible to deal separately with each one.

The term "practice" has become widely used in radiological protection and denotes an activity that causes an increase in exposure to radiation or in the risk of exposure to radiation. A practice can be a business, trade, industry or similar activity. It can also be a government undertaking or a charity. Regardless of its purpose, however, it is implicit in the concept of a practice that the radiation sources that it introduces or maintains can be controlled directly by action on that source.

The exposure of people to ionizing radiation can also be categorized in different ways:

(i) *Medical* exposure of patients, which includes radiation exposure resulting from diagnostic, interventional and therapeutic procedures.

(ii) *Occupational* exposure, in which radiation exposure is incurred as a result of work.

(iii) *Public* exposure, which includes all exposures of the public other than occupational or medical exposure of patients, and includes exposures of the embryo and foetus of pregnant workers.

Of course any particular individual could belong simultaneously to all three categories.

"Patients" are defined as individuals who receive an exposure to radiation associated with a diagnostic, interventional or therapeutic procedure. Dose limits and dose constraints do not apply to individual patients because they may reduce the effectiveness of the patient's diagnosis or treatment, thereby doing more harm than good. Emphasis is therefore placed on the justification of the medical procedures and on the optimisation of protection and, for diagnostic procedures, the use of diagnostic reference levels.

"Workers" are defined as any person who is employed, and who has recognised rights and duties in relation to occupational radiological protection.

Workers in medical professions involving radiation are occupationally exposed, and air crew may also be considered to lie in this category – but not "frequent fliers". (Exceptional cases of cosmic radiation exposures, such as exposure in space travel, where doses may be significant and some type of control warranted, are dealt with separately, taking into account the special type of situations that can give rise to this type of exposure).

An important function of an "employer" is that of maintaining control over the sources of exposure, and over the protection of workers who are occupationally exposed. In this respect, the classification of areas of work is preferable to the classification of workers. There are usually two types of designation – "controlled areas" and "supervised areas". A "controlled area" is a defined area in which specific protection measures and safety provisions are, or could be, required for controlling normal exposures or preventing the spread of contamination during normal working conditions, and preventing or limiting the extent of potential exposures. A "supervised area" is one in which the working conditions are kept under review but for which special procedures are not normally needed. A controlled area is often within a supervised area, but need not be. Workers in controlled areas of workplaces are of necessity well informed and specially trained, and form a readily identifiable group. Such workers are monitored for radiation exposures incurred in the workplace, and occasionally may receive special medical surveillance.

Particular attention is paid to pregnant workers and breast feeding mothers. If a female worker has declared that she is pregnant, additional controls have to be considered to protect the embryo or foetus, to a level that is equivalent to that provided for members of the public. The working conditions of a pregnant worker should therefore be such as to ensure that the additional dose to the embryo or foetus would not exceed about 1 mSv during the remainder of the pregnancy. The principal implication is that the employer should carefully review the exposure conditions of pregnant women and, if required, alter their working conditions so that the probability of accidental doses and radionuclide intakes is extremely low.[16,17]

Finally, a "member of the public" is defined as any individual who receives an exposure that is neither occupational nor medical. A large range of different natural and man-made sources contribute to the exposure of members of the public but, in general, each source will result in a distribution of doses over many individuals. For the purposes of protection of the public, the term "critical group" has long been used to characterise an individual receiving a dose that is representative of the more highly exposed individuals in the population. Dose restrictions were then applied to the mean dose in the appropriate critical group. A considerable body of experience has now been gained in the application of the critical group concept, particularly in the UK. There have also been developments in the techniques used to assess doses to members of the public, including the increasing use of probabilistic techniques. The adjective "critical" has also had the connotation of a "crisis", which was never intended by ICRP. Furthermore, the word "group" can be confusing in the context where the assessed dose is to an individual.

So the ICRP now recommends the use of the "Representative Person" for the purpose of radiological protection of the public instead of the earlier critical group concept.[18] This Representative Person may be real or hypothetical, but it is important that the habits (*e.g.*, consumption of foodstuffs, breathing rate, location, usage of local resources *etc.*) used to characterise the Representative Person are typical habits of a small number of individuals representative of those most highly exposed, and not the extreme habits of a single member of the population. Thus although consideration may be given to some extreme or unusual habits, they should not dictate the characteristics of the Representative Persons considered. Dose coefficients are available for the calculation of prospective doses to different age categories, but for practical reasons it is now recommended that three age categories be used: 0–5 years (infant); 6–15 years (child); and 16–70 years (adult), the dose coefficients and habit data for a 1 year old, a 10 year old, and an adult being used respectively.

All of these concepts and definitions need marshalling together in order to provide advice that is both consistent and logical across all exposure situations, and across all categories of exposure. In order to do so, it is necessary to construct some form of principled framework. Such a framework obviously needs to be based on the scientific information that exists, and the LNT model, whilst also allowing for the incorporation and interpretation of new information as it arises. However, it also needs to be able to accommodate other factors relating to sociological, financial and other relevant considerations if it is to be of value as a decision making tool. The ICRP has attempted to rise to this challenge by basing its advice on the following three *key principles*:

(i) The *Principle of Justification*: any decision that alters the radiation exposure situation should do more good than harm.
(ii) The *Principle of Optimisation of Protection*: the likelihood of incurring exposure, the number of people exposed and the magnitude of their individual doses should all be kept as low as reasonably achievable, taking into account economic and societal factors.
(iii) The *Principle of Application of Dose Limits*: the total dose to any individual from regulated sources in planned exposure situations, other than medical exposure of patients, should not exceed the appropriate limits specified by the ICRP.

The principles of justification and optimisation apply in all three exposure situations, whereas the principle of application of dose limits applies only to doses expected to be incurred with certainty as a result of planned exposure situations.

4.1 Justification

There are two different approaches to applying the principle of justification in situations involving occupational and public exposures, which depend upon whether or not the source can be directly controlled. The first approach is used in the introduction of new activities where radiological protection is planned in advance and the necessary actions can be taken in relation to the source. Application of the justification principle to these situations requires that no planned exposure situation should be introduced unless it produces sufficient net benefit to the exposed individuals, or to society, to offset the radiation detriment it is expected to cause. The second approach is used where exposures can be controlled primarily by action to modify the *pathways* of exposure, and not by acting directly on the source. This is likely to be the case in existing and emergency exposure situations. In these circumstances, the principle of justification is applied when making the decision as to whether or not to take action to avert further exposure. Any decision taken to reduce doses – which will almost always have some disadvantages – should also be justified, in the sense that they should do more good than harm.

In both approaches, the responsibility for judging the "justification" usually falls on governments, or national authorities, to ensure an overall benefit in the broadest sense to society and thus not necessarily to each individual. However, input to the justification decision may include many aspects that could be informed by users or other organisations, or persons outside of government. As such, justification decisions are often informed by some form of public consultation, depending upon, among other things, the size of the source concerned. There are many aspects of justification, and different organisations may be involved and responsible. In this context, radiological protection considerations serve only as one input to the broader decision process.

4.2 Optimisation

The process of optimisation of protection is intended for application to those situations that have been deemed to be justified in the first place. The principle, with restriction on the magnitude of individual dose or risk, is central to the system of protection and applies to all three exposure situations. It is defined by the ICRP as the source-related process to keep the *likelihood* of incurring exposures (where these are not certain to be received), the number of people exposed, and the magnitude of individual doses as low as reasonably achievable, taking economic and societal factors into account. This process of optimisation over several decades has resulted in substantial reductions of occupational and public exposures, and is key to the entire approach currently advocated for radiological protection. Essentially, it is always aimed at achieving the best level of protection under the prevailing circumstances through an ongoing, iterative process that involves:

(i) the evaluation of the exposure situation, including any potential exposures (the framing of the process);
(ii) the selection of an appropriate value for the constraint or reference level;
(iii) the identification of the possible protection options;
(iv) the selection of the best option under the prevailing circumstances; and
(v) the implementation of the selected option.

4.3 Dose Limits

Dose limits only apply to planned exposure situations (but obviously not to medical exposures of patients). For occupational exposure in planned exposure situations, the limit is expressed as an effective dose of 20 mSv per year, averaged over defined five year periods (in other words, 100 mSv in five years), with the further provision that the effective dose should not exceed 50 mSv in any single year. For public exposure in planned exposure situations, the limit is expressed as an effective dose of 1 mSv in a year, but in special circumstances a higher value could be allowed in a single year, provided that the average, again over defined five year periods, does not exceed 1 mSv per year.

The limits on effective dose apply to the sum of doses due to both external exposures and to committed doses from internal exposures arising from the intake of radionuclides. Occupational intakes may be averaged over a period of five years to provide some flexibility. Similarly, the averaging of public intakes over a period of five years would be acceptable in those special circumstances where averaging of the dose to members of the public could be allowed.

Dose limits do not apply in emergency exposure situations where an informed, exposed individual is engaged in volunteered life-saving actions, or is attempting to prevent a catastrophic situation. For informed volunteers undertaking urgent rescue operations, the normal dose restriction may be relaxed. However, responders undertaking recovery and restoration operations in a later phase of emergency exposure situations should be considered as occupationally exposed workers, and thus protected according to normal occupational radiological protection standards, and their exposures should not exceed the occupational dose limits. (Female workers who are pregnant, or are nursing an infant, should not be employed as "first responders" undertaking life-saving or other urgent actions).

Notwithstanding the basic scientific method adopted by the ICRP, its approach to the selection of dose limits necessarily includes societal judgments applied to the many and varied attributes of "risk". Not only would these judgments probably be different from one operational context to another within any given society, but they are also likely to differ from one society to another. Providing general guidance is therefore not that easy, and the ICRP makes it clear that it is for this reason that its guidance is intended to be sufficiently flexible to allow for national or regional variations.

4.4 Dose Constraints and Reference Levels

The concepts of dose constraints and reference levels are used in conjunction with the optimisation of protection to restrict individual doses. A level of individual dose, either as a dose constraint or a reference level, always needs to be defined. Thus the initial *intention* would be to not exceed, or remain at, these levels; and the *ambition* would be to reduce all doses to levels that are as low as reasonably achievable, economic and societal factors being taken into account. In planned exposure situations (with the exception of medical exposure of patients), the term used is "dose constraint", but for emergency and existing exposure situations the term "reference level" is used to describe this level of dose. The difference in terminology is because that, in planned situations, the restriction on individual doses can be applied at the planning stage, and the doses can be forecast so as to ensure that the constraint will not be exceeded. In the other situations, however, a wider range of exposures may exist, and the optimisation process may apply to initial levels of individual dose above the reference level. [Diagnostic reference levels are used in medical diagnosis (*i.e.*, planned exposure situations) to indicate whether, in routine conditions, the levels of patient dose or administered activity from a specified imaging procedure are unusually high or low for that procedure].

A dose constraint is thus a prospective and *source-related* restriction on the individual dose from a source in planned exposure situations, serving as an upper bound on the predicted dose in the optimisation of protection for that particular source. It is a level of dose above which it is unlikely that protection is optimised for a given source of exposure, and will always be lower than the pertinent dose limit. For potential exposures, the corresponding source-related restriction is called a "risk constraint".

For occupational exposures, the dose constraint is a value of individual dose used to limit the range of options, such that only options expected to cause doses below the constraint are considered in the process of optimisation. For public exposures, the dose constraint is an upper bound on the annual doses that members of the public could receive from the planned operation of a specified controlled source. If a dose constraint is exceeded, then it is necessary to determine whether protection had been optimised, the appropriate dose constraint had been selected, and further steps to reduce doses to acceptable levels would have been appropriate.

Emergency or existing controllable exposure situations are somewhat different. Here the reference levels represent the level of dose or risk above which it is judged to be inappropriate to plan to allow exposures to occur, and for which therefore protective actions should be planned and optimised. The chosen value for a reference level will depend upon the prevailing circumstances of the exposure situation. Quite obviously, when an emergency exposure situation has occurred, or an existing exposure situation has been identified, and protective actions have been implemented, doses to workers and members of the public can be measured or assessed. The reference level may then assume a different function, and serve essentially as a benchmark against which protection options can be judged retrospectively. One has to bear in mind that the distribution of doses that result from the implementation of a planned protective strategy may or may not include exposures above the reference level, depending on the success of the strategy.

Because at doses higher than 100 mSv there is an increased likelihood of deterministic effects, and a significant risk of cancer, the maximum value for a reference level should be 100 mSv incurred either acutely or in a year. Higher exposures would only be justified under extreme circumstances, and thus either because the exposure is unavoidable or because the situation was exceptional – such as the saving of life or the prevention of a serious disaster. No other individual or societal benefit would compensate for such high exposures.

Below this, the values fall into three defined bands. They apply across all three exposure situations and refer to the projected dose over a time period that is appropriate for the situation under consideration. Constraints for planned exposures and reference levels in existing situations are conventionally expressed as an annual effective dose (mSv in a year). In emergency situations, the reference level will be expressed as the total residual dose to an individual as a result of the emergency that the regulator would plan not to exceed, either acute (and not expected to be repeated) or, in case of protracted exposure, on an annual basis.

The first band, of 1 mSv or less, applies to exposure situations where individuals receive exposures – usually planned – that may be of no direct benefit to them, but the exposure situation may be of benefit to society, the exposure of members of the public from the planned operation of nuclear power being a prime example. Constraints and reference levels in this band would be selected for situations where there is general information and environmental surveillance, or monitoring, or assessment, and where individuals may receive information but no training. The corresponding doses would represent a marginal increase above the natural background, and at least two orders of magnitude lower than the maximum value for a reference level, thus providing a rigorous level of protection.

The second band, greater than 1 mSv but not more than 20 mSv, applies in circumstances where individuals receive direct benefits from an exposure situation. Constraints and reference levels in this band will often be set in circumstances where there is individual surveillance or dose monitoring or assessment, and where individuals benefit from training or information – as is the case in occupational exposure from planned exposure situations. Abnormally high levels of natural background radiation, or stages in post-accident rehabilitation, may also be in this band.

The third band, greater than 20 mSv but not more than 100 mSv, applies in unusual, and often extreme, situations where actions taken to reduce exposures would be disruptive. Reference levels and, occasionally for "one-off" exposures below 50 mSv, constraints could also be set in this range in circumstances where benefits from the exposure situation are commensurately high. Action taken to reduce exposures in a radiological emergency is the main example of this type of situation. Any dose rising towards 100 mSv will almost always justify protective action. In addition, situations in which the dose threshold for deterministic effects in relevant organs or tissues could be exceeded should always require action.

5 Radiation Protection in Practice in the UK

So how is all of this extensive advice and guidance supposed to be implemented and actually applied to different radiation exposure situations? It goes without saying that the detailed interpretation can be complex indeed, but by and large it all follows much the same pattern, which can be summarized as follows:

(i) A characterisation of the possible situations where radiation exposure may occur (planned, emergency and existing exposure situations).

(ii) A classification of the types of exposure (those that are certain to occur and potential exposures, as well as occupational exposure, medical exposure of patients and public exposure).
(iii) An identification of the exposed individuals (workers, patients and members of the public).
(iv) A categorisation of the types of assessment, namely source-related and individual-related.
(v) A precise formulation of the principles of protection: justification, optimisation of protection, and the application of dose limits.
(vi) A description of the levels of individual doses that require protective action or assessment (dose limits, dose constraints and reference levels).
(vii) A delineation of the conditions for the safety of radiation sources, including their security and the requirements for emergency preparedness and response.

Radiological protection is undertaken within the UK under various pieces of legislation, including a number of regulations, the developments of which have been recorded in some detail by O'Riordan.[19]

5.1 Radiation Exposure of Workers

The principal doses received by workers are associated with exposure to X-rays and gamma rays together with, but to a lesser extent, beta particles and neutrons. Doses at the body surface are usually estimated by the use of personal dosimeters, but assessments are also made in relation to internal exposures where relevant. Measurements are usually made to ensure compliance with legal or administrative dose limits. Workers employed by the nuclear industries are involved in many different aspects of the nuclear fuel cycle; fuel fabrication; reactor operation; the care, maintenance and decommissioning of reactors; and the restoration of nuclear sites. A variety of other workers may be exposed to radiation sources in the course of their work, as well as those involved in military defence and its associated industries, and those in nuclear medicine and radiography.

Annual dose rates received by nuclear workers are typically of a few mSv or less. What is of particular interest, however, is their long term collective dose, and how this relates to their general health, and particularly to the incidence of cancer amongst them. Several studies have been made, the most recent being that conducted by the Health Protection Agency.[20] This study involved the data from some 174 500 workers from pre-1976 (back to the end of the Second World War) up to 2001. About 68% of them had a lifetime dose of 10mSv or less, about 20% had a lifetime dose of 10 to 50 mSv, 6% lifetime doses up to 100 mSv, and the remaining 6% lifetime doses in excess of 100 mSv.

The results of the study showed that, as in previous analyses, total mortality and mortality from major causes were less than expected from rates for England and Wales. This "healthy worker effect" remains even after adjustment for social class. The only cause for which mortality was statistically significantly greater than that expected from national rates was pleural cancer, and this probably reflected exposure to asbestos.

Mortality and incidence from both leukaemia (excluding chronic lymphatic leukaemia), and the grouping of all malignant neoplasms other than leukaemia, increased to a statistically significant extent with increasing external radiation dose. The corresponding central estimates of the trend in risk with dose were similar to those for the survivors of the atomic bombings of Hiroshima and Nagasaki. And whilst there was some evidence of an increasing trend with dose in mortality from all circulatory diseases combined, the irregular pattern in risk with dose and similarities with the corresponding pattern for lung cancer suggest that this finding may, at least in part, be due to confounding by smoking. In contrast, both for mortality and incidence, the trend with dose in the risk of all malignant neoplasms (other than leukaemia) would not appear to be an artifact due to smoking, because the relationship remanis the same if the data for lung and pleural cancer (which are related to smoking) are excluded.

5.2 *Radiation Exposure of the Public*

In recent years, radioactivity in the environment has come from several sources. These include natural radiation, residues from the Chernobyl accident and from the atmospheric testing of weapons, plus radioactive discharges and emissions from nuclear and non-nuclear sites (so-called "authorised premises"). Nuclear licensed sites, which are subject to the Nuclear Installations Act[21] may also be authorised to dispose of radioactive wastes under the Radioactive Substances Act.[22] These discharges are primarily liquid, and made into rivers, estuaries or coastal waters. Discharges of radioactive wastes from other sites, such as hospitals, industrial sites and research establishments, are also regulated under this Act but are not subject to the Nuclear Installations Act. Small amounts of very low level solid radioactive waste are routinely disposed of from some non-nuclear sites, and there is also a significant radiological impact due to the legacy of past discharges of radionuclides from non-nuclear industrial activity in the UK. These involve radionuclides that also occur naturally in the environment. Discharges from terrestrial non-nuclear sites are generally considered insignificant, and as such environmental monitoring of their effects is usually not required for the purposes of protection of public health in the UK. This situation is, however, reviewed from time to time.

The discharge limits are set through an authorisation assessment process which can be initiated by either the operator or the relevant environment agency. In support of the assessment process, prospective assessments of doses to the public are made assuming discharges are kept within the authorised limits. Authorisations are then set so that doses to the public from the site will be below the dose constraint of 0.3 mSv per year (or 0.5 mSv per year if discharges occurred actually at the authorised limits) for that source – the dose limit for the public from all sources being 1 mSv per year.

The Environment Agencies set limits and regulate the discharges and emissions of radioactive waste from authorised premises. Operators of nuclear sites are required both to monitor their discharges and the effects on the environment. In England, Wales and Northern Ireland, the Food Standards Agency, the Environment Agency and the Northern Ireland Environment Agency conduct their own monitoring programmes, whereas in Scotland the Scottish Environment Protection Agency incorporates the requirements of the Food Standards Agency within its own programme. These programmes are important because they provide an independent assessment of the potential harm resulting from authorised releases of radioactive discharges, and act as an additional check to the monitoring programmes conducted by site operators.

The assessments are based on a collection of data relating to the radionuclide concentrations of foodstuffs, external dose rates and information on the habits of people living near the sites. Changes in doses received do occur from year to year, usually because of variations in concentrations of radionuclides in food and in the external dose rates, but in some years doses are affected by changes in people's habits, in particular their consumption of food, which are identified by carrying out regular food habits' surveys.

In recent years, a group of people in Cumbria that consume a large amount of fish and shellfish have received the highest dose of radiation due to discharges from two different sources. Their dose was estimated to be 0.52 mSv in 2007 (ref. 23). This was due to the effects of authorized current and past liquid discharges from the reprocessing plant at Sellafield into the Irish Sea, and from past liquid discharges from a phosphate processing plant at Whitehaven a few miles up the coast from Sellafield. The Sellafield discharges were estimated to have contributed 0.24 mSv to this dose in 2007, primarily due to the accumulation of caesium-137, plutonium isotopes and americium-241 in seafood, from past liquid discharges, as well as external exposure from contaminated sediment. The phosphate plant's discharges (of what are known as "technologically enhanced naturally-occurring radioactive material", where there is an increase in concentrations of some radionuclides that occur naturally due to industrial operations) resulted in the people who consumed seafood also receiving 0.28 mSv from that source. This was due to polonium-210 concentrations in seafood, which occur naturally anyway, but which also partly arise from the decay of radium-226 and lead-210 in past discharges from the phosphate plant.

Doses to people who had consumed crops grown on land fertilised by seaweed from around Sellafield were also assessed and their estimated dose for 2007 was 0.012 mSv. Doses to people using the beaches and other intertidal areas in the vicinity were less than 0.02 mSv.

People living around operating nuclear reactors generating electricity within the UK receive doses that are typically less than 0.1 mSv per year in 2007. Such low doses often then raise the question of how they compare with natural background radiation. In doing so, however, it must be recalled that radiological protection is based on the premise that an increment in dose results in an increase in risk, the increment being on top of whatever the existing dose rate may be. Nevertheless, it is sometimes useful to note that the background dose rate in the UK can typically vary from about 1.5 to 7.5 mSv per year, with an average of about 2.2 mSv per year, the variation being due primarily to radon. Where radon exposure levels in homes are high, action to reduce them is encouraged. For comparison, it may also be noted that a typical single chest X-ray would give a dose of about 0.02 mSv and a chest CT scan about 8 mSv.

6 Experience Gained from Nuclear Accidents Outside the UK

All of the situations discussed in the preceding section apply to planned 'normal' exposures. On rare occasions, however, abnormal situations arise, the most recent being that at Fukushima in Japan. This happened some twenty five years after the Chernobyl accident, which occurred on 26 April 1986 during a low power engineering test of their Unit 4 reactor. At the time of writing, little is known about the events at Fukushima, but it is useful to review what has been learned from Chernobyl.

The Chernobyl site is located in present-day Northern Ukraine, some 20 km south of the border with Belarus and 140 km west of the border with the Russian Federation. The accident was caused by the improper operation of the reactor, which itself had severe design flaws, allowing an uncontrollable power surge to occur. This resulted in successive explosions that severely damaged the reactor building and completely destroyed the reactor. The accident caused the uncontrolled release of large quantities of radioactive substances into the air for about 10 days. The radioactive cloud dispersed over the entire northern hemisphere and deposited substantial amounts of radioactive material. At the site itself, two workers died from injuries, and approximately 600 workers responded within the first day to the immediate emergency, including staff at the plant, firemen, security guards, and staff of the local medical facility. The dominant exposures for these personnel were external irradiation of the whole body at high dose rates, and beta-irradiation of the skin. Internal contamination was of relatively minor importance, and neutron exposure was insignificant. As was to be expected, a very considerable effort has since been expended to follow up on the human consequences of this major nuclear disaster, and the latest findings are those of UNSCEAR.[24]

Cases of acute radiation syndrome (ARS) occurred among the plant employees and so-called 'first responders' but not among the evacuated populations or the general population. The diagnosis of ARS was initially considered for 237 persons, based on symptoms of nausea, vomiting and diarrhoea. The diagnosis was confirmed in 134 persons. There were 28 early deaths (first four months), primarily (95%) where whole-body doses were in excess of 6.5 Gy. Underlying bone marrow failure was the main contributor to all deaths during the first two months, in spite of attempts to save them with bone marrow transplants. Skin doses exceeded bone marrow doses by a factor of 10 to 30, and many ARS patients received skin doses in the range of 400–500 Gy. Radiation damage to the skin aggravated other conditions, and this was considered to be a major contributor to at least 19 of the deaths. Such damage significantly increased the severity of the ARS, especially when skin burns exceeded 50% of the body surface area and led to major infections. Since then, 19 ARS survivors have died (up to 2006), but their deaths have been attributed to various causes, and usually not associated with radiation exposure. Skin injuries and radiation-induced cataracts are, however, major lasting clinical impacts for the ARS survivors.

In 1986 and 1987, some 440 000 recovery operation workers were used at the Chernobyl site, and more 'recovery workers' were involved in various activities between 1988–1990. Collectively, about 600 000 persons (civilian and military) received special certificates confirming their status as recovery operation workers (unfortunately also known as "liquidators"). About 240 000 were military servicemen. The average effective dose received by these recovery operation workers between 1986–1990, and mainly due to external irradiation, is estimated to have been about 120 mSv. The recorded worker doses varied from $>$ 10 mSv to $<$ 1 Sv, although about 85% of the recorded doses were in the range 20–500 mSv. (Uncertainties in the individual dose estimates vary from $>$ 50% up to a factor of 5, and the estimates for the military personnel are suspected to be biased towards high values.) To date, there is some evidence of a detectable increase in the incidence of leukemia, primarily based upon results from the Russian Federation, and of cataracts among those who received higher doses, but there is no evidence of other health effects than can be attributed to radiation exposure.

With regard to the public, the number of evacuees was about 115 000, consisting of about 25 000 persons from Belarus, 200 from the Russian Federation and 90 000 from the Ukraine. The areas from which people were evacuated form what is called the "exclusion zone", which includes not only the 30 km zone, which is the area within a 30 km radius centred on the location of the Chernobyl reactor, but also highly-contaminated areas adjacent to the 30 km zone and more distant areas where high levels of radionuclide deposition density were measured.

Two radionuclides, the short-lived iodine-131 (with a half-life of 8 days) and the longer-lived caesium-137 (with a half-life of 30 years), were particularly significant for the radiation dose they delivered to members of the public. In the former Soviet Union the contamination of fresh milk with iodine-131, and the lack of prompt countermeasures, led to high thyroid doses, particularly among children. The thyroid doses received by the evacuees varied according to their age, place of residence, consumption habits, and date of evacuation. For many pre-school children the doses to the thyroid were well in excess of 1 Gy. It is therefore not surprising that there has been a substantial increase in thyroid cancer incidence amongst those exposed as children or adolescents in Belarus, the Russian Federation, and the Ukraine since the Chernobyl accident, and this increase has shown no signs of diminishing (up to 20 years after exposure). Amongst those under the age of 14 years in 1986, 5127 cases (for those under the age of 18 years in 1986, 6848 cases) of thyroid cancer have been reported between 1991–2005 for the whole of Belarus and Ukraine and the four more affected regions of the Russian Federation. By 2005, 15 cases had proved fatal.

In the longer term, mainly due to caesium-137, the general population was also exposed to radiation externally from radioactive deposition and internally from consuming contaminated foodstuffs. The resulting radiation doses were relatively low, however, partly because of the countermeasures taken. Excluding doses to the thyroid, the mean effective doses due to external irradiation were estimated to have been about 30 mSv for the Belarusian evacuees, about 25 mSv for the Russian evacuees, and about 20 mSv for the Ukrainian evacuees. These values were at least 10 times smaller than the corresponding numerical values of thyroid doses resulting from internal irradiation. The mean effective doses due to internal irradiation were estimated to have been about 6 mSv for the Belarusian evacuees, about 10 mSv for the Ukrainian evacuees, and about 10 mSv for the Russian evacuees. These values were at least half of the corresponding effective doses due to external irradiation.

Among those exposed *in utero* and as children, no persuasive evidence has apparently accrued to suggest that there is a measurable increase in the risk of leukemia due to radiation exposure. This is not unreasonable, because the doses involved were generally very small, and therefore epidemiological studies would lack sufficient statistical power to observe any effect, had there been one. Overall, therefore, the average effective doses, due to both external and internal exposures, received by members of the public during 1986–2005 were about 30 mSv for the evacuees, 1 mSv for the residents of the former Soviet Union, and 0.3 mSv for the populations of the rest of Europe.

More recently, on 11 March 2011, an earthquake and accompanying tsunami struck the coastal area of Japan and caused major damage to the Fukushima Dai-ichi nuclear power plant, which consists of six boiling water reactors, three of which were operating at the time. Three staff were killed as a result of these events – not related to radiation exposure. When the earthquake struck the reactors automatically shut down and the emergency cooling systems were activated but one hour later these were all damaged by a wall of water some 14 m high as a result of the accompanying tsunami. (The tsunami itself was responsible for the deaths of over 26 000 local residents.) Hydrogen explosions subsequently badly damaged the control rooms of the three operating reactors (Units 1, 2 and 3) and there were problems with the spent fuel pool of Unit 4, which subsequently led to a fourth hydrogen explosion. There have been no recorded cases of ARS amongst the staff dealing with the emergency, and none are expected.

Local residents were evacuated out of the area in a staged manner up to a radius of 20 km around the site, the evacuation being compounded by evacuees from the tsunami. The principal nuclides of concern were again those of iodine and caesium. Residents within a 20–30 km radius were instructed to shelter indoors. In contrast to Chernobyl, the radionuclides released were not widely distributed and considerable precipitation subsequently occurred due to snowfall. Protective actions were immediately implemented with regard to the consumption of contaminated water and foodstuffs and the screening of children, in particular, for iodine concentrations in the thyroid gland was undertaken. More detailed information is still awaited but clearly the major long-term impact for the local population, having experienced a severe earthquake, tsunami, and a nuclear accident, will be psychological.

7 Conclusions

The current system of radiological protection for people has been developed over a long period of time, and has involved an enormous body of scientific, medical and cultural information. All of these areas are still actively pursued, and the system reviewed and revised. In terms of application, an enormous amount of experience has now been gathered over many decades. Exposures of people to ionizing radiation may be through medical diagnostic or therapeutic exposures, of which there must be a vast number undertaken daily throughout the world; through exposures at work in all forms of industry that may involve radioactive or radiation sources; or through public exposures arising from releases from both nuclear and non-nuclear establishments. All of these exposures, and the sources leading to them, are controlled on the same scientific basis and interpretation, and on the advice of the same international committee-the ICRP, and its extensive support. If there was something seriously amiss with this system, then it would by now have come to light. Not that there is any reason to be complacent, as the recent incident at Fukushima, and the 25th anniversary of Chernobyl serve to remind us, accidents can happen, as they can in any industrial endeavor. But it should provide a high degree of assurance to anyone that is concerned about radiation safety and our ability to manage it safely that, whatever the source of exposure or the category of people exposed, the actions taken to safeguard human health are based on a wealth of experience that is unequalled in any other field.

References

1. ICRP, *1959 Recommendations of the International Commission on Radiological Protection*, ICRP Publication 1, Pergamon Press, Oxford, UK, 1959.
2. ICRP, *1977 Recommendations of the International Commission on Radiological Protection*, ICRP Publication 26. Ann. ICRP 1(3), 1977.
3. ICRP, *1991 Recommendations of the International Commission on Radiological Protection*, ICRP Publication 60, Ann. ICRP 21(1–3), 1991.
4. ICRP, *2007 Recommendations of the International Commission on Radiological Protection*, ICRP Publication 103, Ann. ICRP 37(2–4), 2007.
5. ICRU, 1938 Recommendations of the International Commission on Radiation Units, Chicago, *Am. J. Roentgenol., Radium Therapy Nucl. Med*, 1937, **39**, 295.
6. ICRU, 1954 Recommendations of the International Commission on Radiation Units, Copenhagen, *Radiology*, 1953, **62**, 106.
7. ICRU, 1962 Radiation Quantities and Units, Report 10a of the International Commission on Radiation Units and Measurements, *Natl. Bur. Std. Handbook*, 1962, 78.
8. ICRP, *2002 Basic Anatomical and Physiological Data for Use in Radiological Protection*, ICRP Publication 89, Ann. ICRP 32(3–4), 2002.
9. R. J. Pentreath, Radioecology, radiobiology, and radiological protection: frameworks and fractures, *J. Environ. Radioct.*, 2009, **100**, 1019–1026.

10. UNSCEAR, *Sources and Effects of Ionizing Radiation*, United Nations Scientific Committee on the Effects of Atomic Radiation Report to the General Assembly with Scientific Annexes, United Nations, New York, NY, USA, 2000, vol. 2 (Effects).
11. NCRP, *Evaluation of the Linear-Non Threshold Dose–Response Model for Ionizing Radiation*, NCRP Report No. 136, National Council on Radiation Protection and Measurements, Bethesda, MD, USA, 2001.
12. NAS/NRC, *Health Risks from Exposure to Low Levels of Ionizing Radiation: BEIR VII Phase 2*, Board on Radiation Effects Research, National Research Council of the National Academies, Washington, DC, USA, 2006.
13. French Academies Report, *La Relation Dose–Effet et l'Estimation des Effets Cancérogènes des Faibles Doses de Rayonnements Ionisants*, 2005, http://www.academiesciences.fr/publications/rapports/pdf/dose_effet_07_04_05.pdf.
14. C. Streffer, H. Bolt, D. Follesdal *et al.*, *Low Dose Exposures in the Environment: Dose-Effect Relations and Risk Evaluation*, Wissenschaftsethik und Technikfolgenbeurteilung, Band 23, Springer, Berlin, Germany, 2004.
15. ICRP, *2005 Protecting People against Radiation Exposure in the Event of a Radiological Attack*, ICRP Publication 96, Ann. ICRP 35(1), 2005.
16. ICRP, *2000 Pregnancy and Medical Radiation*, ICRP Publication 84, Ann. ICRP 30(1), 2000.
17. ICRP, *2001 Doses to the Embryo and Embryo/Fetus from Intakes of Radionuclides by the Mother*, ICRP Publication 88, Ann. ICRP 31(1–3), 2001.
18. ICRP, *2006 Assessing Dose of the Representative Person for the Purpose of Radiation Protection of the Public and the Optimisation of Radiological Protection: Broadening the Process*, ICRP Publication 101, Ann. ICRP 36(3), 2006.
19. M. O'Riordan, *Radiation Protection: A Memoir of the National Radiological Protection Board*, Health Protection Agency, UK, 2007.
20. C. R. Muirhead, J. A. O'Hagan, R. G. E. Haylock, M. A. Phillipson, T. Willcock, G. L. C. Berridge and W. Zhang, Health Protection Agency, UK, 2009.
21. UK Parliament, *1965 Nuclear Installations Act*, HMSO, London, UK.
22. UK Parliament, *1993 Radioactive Substances Act*, HMSO, London, UK.
23. RIFE, *2008 Radioactivity in Food and the Environment*, Environment Agency, Food Standards Agency, Northern Ireland Environment Agency, Scottish Environment Protection Agency, UK, 2007.
24. UNSCEAR, *Sources and Effects of Ionizing Radiation, UNSCEAR 2008 Report to the General Assembly with Scientific Annexes, Vol. II, Annex D, Health Effects due to Radiation from the Chernobyl Accident,* United Nations, New York, 2011.

Subject Index

Lightning Source UK Ltd.
Milton Keynes UK
UKOW07n1813301114

242422UK00002B/78/P

9 781849 731942